# Learning ArcGIS for Desktop

Create, analyze, and map your spatial data with ArcGIS for Desktop

**Daniela Cristiana Docan**

PUBLISHING

BIRMINGHAM - MUMBAI

# Learning ArcGIS for Desktop

First published: March 2016

Production reference: 1220316

Published by Packt Publishing Ltd.
Livery Place
35 Livery Street
Birmingham B3 2PB, UK.

ISBN 978-1-78217-579-7

www.packtpub.com

Cover image by Emilian Dobroslav (emilian18@gmail.com)

# Credits

**Author**
Daniela Cristiana Docan

**Reviewer**
Tripp Corbin, GISP

**Commissioning Editor**
Dipika Gaonkar

**Acquisition Editor**
Vivek Anantharaman

**Content Development Editor**
Anish Dhurat

**Technical Editor**
Hussain Kanchwala

**Copy Editors**
Priyanka Ravi
Sonia Mathur

**Project Coordinator**
Bijal Patel

**Proofreader**
Safis Editing

**Indexer**
Rekha Nair

**Production Coordinator**
Manu Joseph

**Cover Work**
Manu Joseph

# About the Author

**Daniela Cristiana Docan** is currently a lecturer at the Department of Topography and Cadastre at the Faculty of Geodesy in Bucharest, Romania. She obtained her PhD in 2009 from the Technical University of Civil Engineering, Bucharest, with her thesis *Contributions to quality improvement of spatial data in GIS*. Formerly, she worked at Esri Romania and National Agency for Cadastre and Land Registration (ANCPI).

While working for Esri Romania, she trained teams (as an authorized instructor in ArcGIS for Desktop by Esri) from state- and privately-owned companies, such as the Romanian Aeronautical Authority, the Agency of Payments and Intervention for Agriculture (APIA), and the Institute of Hydroelectric Studies and Design. She also trained and assisted the team in charge of quality data control in the Land Parcel Identification System (LPIS) project, in Romania.

For the ANCPI, she created the logical and physical data model for the Romanian National Topographic Dataset at a scale of 1:5,000 (TOPRO5) in 2009. She was a member of the workgroup that elaborated TOPRO5 and its metadata technical specifications and the *Report on the implementation of the INSPIRE Directive* in Romania in 2010.

Prior to this book, Daniela worked on *ArcGIS for Desktop Cookbook*, *Packt Publishing*, which covers the following topics: designing a file geodatabase schema, constraining the geometry and attribute values of the data, geocoding addresses, working with routes and events, and using spatial ETL tools.

I would like to thank Mohammed Fahad, Anish Dhurat, Hussain Kanchwala, and everyone else at Packt Publishing who worked on this book.

A special thanks to technical reviewer, Tripp Corbin. His work and practical advice made this book better.

I want to express my gratitude to Emilian and my friends for their continuous support.

# About the Reviewer

**Tripp Corbin, GISP** is the CEO and a cofounder of eGIS Associates, Inc. He has over 20 years of surveying, mapping, and GIS-related experience. He is recognized as an industry expert with a variety of geospatial software packages, including Esri, Autodesk, and Trimble products. He holds multiple certifications, including Microsoft Certified Professional, Certified Floodplain Manager, Certified GIS Professional, CompTIA Certified Technical Trainer, Esri Certified Enterprise System Design Associate, and Esri Certified Desktop Professional.

Tripp is a very active member of the GIS professional community. He currently serves as the President Elect of URISA and as an At-Large GITA Southeast Board Member. In recognition of his contributions to the GIS community, he has received several awards, including the URISA Exemplary Leadership Award and the Barbara Hirsch Special Service Award. Tripp also recently authored the book, *Learning ArcGIS Pro*, *Packt Publishing*, which is the first book published on Esri's newest desktop GIS application ArcGIS Pro.

# www.PacktPub.com

## eBooks, discount offers, and more

Did you know that Packt offers eBook versions of every book published, with PDF and ePub files available? You can upgrade to the eBook version at www.PacktPub.com and as a print book customer, you are entitled to a discount on the eBook copy. Get in touch with us at customercare@packtpub.com for more details.

At www.PacktPub.com, you can also read a collection of free technical articles, sign up for a range of free newsletters and receive exclusive discounts and offers on Packt books and eBooks.

https://www2.packtpub.com/books/subscription/packtlib

Do you need instant solutions to your IT questions? PacktLib is Packt's online digital book library. Here, you can search, access, and read Packt's entire library of books.

## Why subscribe?

- Fully searchable across every book published by Packt
- Copy and paste, print, and bookmark content
- On demand and accessible via a web browser

# Table of Contents

# Preface

Welcome to *Learning ArcGIS for Desktop*. ArcGIS for Desktop is one of the main components of Esri's ArcGIS platform, which is used to support decision making and solve various mapping problems. It contains a wide variety of tools to create, manage, analyze, map, and share spatial data.

*Learning ArcGIS for Desktop* starts with the computer hardware and software recommendations. Then, this book goes on to show you how to obtain and install a 60-day trial of ArcGIS for Desktop (Advanced) on Windows. The second chapter explores coordinate reference system concepts. In the next three chapters, you will learn how to create a file geodatabase and manage, create, edit, and symbolize spatial data. Then, this book focuses on planning and performing spatial analysis on vector data using geoprocessing tools and ModelBuilder. Next, you will analyze raster data using the Spatial Analyst and 3D Analyst extensions. Finally, basic principles of cartography design will be used to create a professional poster map.

The book is a tutorial-based guide that will lead you through the basic concepts and functions of Esri's ArcGIS for Desktop software.

## What this book covers

*Chapter 1, Getting Started with ArcGIS*, covers the hardware and software requirements and shows you how to obtain and install a 60-day trial of ArcGIS for Desktop Advanced, single-use version. This chapter introduces you to the main ArcGIS for Desktop applications: ArcCatalog and ArcMap.

*Chapter 2, Using Geographic Principles*, explains the basic concepts of geographic and projected coordinate systems. You will explore the major categories of map projections using the ArcMap application. Furthermore, you will learn how to use the ArcGIS datum transformations to correctly convert and transform different coordinate reference systems.

*Chapter 3, Creating a Geodatabase and Interpreting Metadata,* shows you how to organize the spatial datasets acquired from external resources in a file geodatabase. You will also learn how to document your file geodatabase using two metadata standards, ISO19139 and INSPIRE.

*Chapter 4, Creating Map Symbology,* shows you how to create and customize symbols and labels on a map. You will learn how to display geographic features based on their attributes using symbols to create qualitative and quantitative thematic maps.

*Chapter 5, Creating and Editing Data,* explains how to create and edit data. You will learn to work with editing tools to create and edit feature shapes and attributes. Also, you will learn how to create point geometry using tabular data.

*Chapter 6, Analyzing Geographic Data and Presenting the Results,* covers how to plan and perform data analysis. You will learn to prepare and combine the spatial datasets to obtain new information using specific analysis tools. Furthermore, you will learn how to generate a report to present the results of your spatial analysis.

*Chapter 7, Working with Geoprocessing Tools and ModelBuilder,* describes the advanced tools to automate an analysis workflow. You will gain a deeper understanding of GIS analysis by working with the geoprocessing tools and models.

*Chapter 8, Using Spatial Analyst and 3D Analyst,* covers how to visualize and analyze vector and raster data using the Spatial Analyst and 3D Analyst extensions. You will learn to perform site selection and a least-cost path analysis using raster data. You will also learn how to create 3D features from 2D features and how to calculate surface area and volume.

*Chapter 9, Working with Aerial and Satellite Imagery,* explains the image-processing functions. You will learn how to georeference an aerial photograph. You will also use the Image Analysis toolbar to display and extract information from the satellite imagery.

*Chapter 10, Designing Maps,* describes the main cartographic design principles that are applied in the ArcGIS Map Layout. You will learn to add, customize, and organize map elements in a map layout. Moreover, you will learn how to create a professional poster map using a standard template from the ArcGIS collection of templates.

# What you need for this book

To complete the exercises in this book, you will need ArcGIS for Desktop 10.3 or 10.4 (Standard or Advanced) installed on your system.

Depending on your software version, please download and install the latest patches (bug fixes) or service packs (compilation of bug fixes) from `http://support.esri.com/en/downloads/patches-servicepacks`.

You need a web browser and access to an Internet connection to add datasets from ArcGIS Online and other public sources.

Data used in this book is freely available on the Packt Publishing site.

# Who this book is for

*Learning ArcGIS for Desktop* is for users who are comfortable with the basic concepts of Geographic Information Systems and want to learn how to create and edit geospatial data, perform spatial analysis, and create effective maps with ArcGIS for Desktop.

# Conventions

In this book, you will find a number of text styles that distinguish between different kinds of information. Here are some examples of these styles and an explanation of their meaning.

Code words in text, database table names, folder names, filenames, file extensions, pathnames, dummy URLs, user input, and Twitter handles are shown as follows: "Start ArcMap application and open your map document named `AccessingImagery.mxd` from `<drive>:\LearningArcGIS\Chapter9\MosaicData`."

When we wish to draw your attention to a particular item, the words are shown as follows: "The result will be a *high resolution multiband image* or a *pan-sharpened multispectral image* with a spatial resolution of 15 meters."

**New terms** and **important words** are shown in bold. Words that you see on the screen, in menus or dialog boxes for example, appear in the text like this: "Use the **Select Features** tool that is located on the **Tools** toolbar to select the five visible city points."

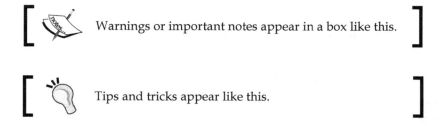

Warnings or important notes appear in a box like this.

Tips and tricks appear like this.

# Reader feedback

Feedback from our readers is always welcome. Let us know what you think about this book—what you liked or disliked. Reader feedback is important for us as it helps us develop titles that you will really get the most out of.

To send us general feedback, simply e-mail feedback@packtpub.com, and mention the book's title in the subject of your message.

If there is a topic that you have expertise in and you are interested in either writing or contributing to a book, see our author guide at www.packtpub.com/authors.

# Customer support

Now that you are the proud owner of a Packt book, we have a number of things to help you to get the most from your purchase.

# Downloading the example code

You can download the example code files for this book from your account at http://www.packtpub.com. If you purchased this book elsewhere, you can visit http://www.packtpub.com/support and register to have the files e-mailed directly to you.

You can download the code files by following these steps:

1. Log in or register to our website using your e-mail address and password.
2. Hover the mouse pointer on the **SUPPORT** tab at the top.
3. Click on **Code Downloads & Errata**.
4. Enter the name of the book in the **Search** box.
5. Select the book for which you're looking to download the code files.
6. Choose from the drop-down menu where you purchased this book from.
7. Click on **Code Download**.

Once the file is downloaded, please make sure that you unzip or extract the folder using the latest version of:

* WinRAR / 7-Zip for Windows
* Zipeg / iZip / UnRarX for Mac
* 7-Zip / PeaZip for Linux

# Downloading the color images of this book

We also provide you with a PDF file that has color images of the screenshots/diagrams used in this book. The color images will help you better understand the changes in the output. You can download this file from `http://www.packtpub.com/sites/default/files/downloads/LearningArcGISforDesktop_ColorImages.pdf`.

# Errata

Although we have taken every care to ensure the accuracy of our content, mistakes do happen. If you find a mistake in one of our books—maybe a mistake in the text or the code—we would be grateful if you could report this to us. By doing so, you can save other readers from frustration and help us improve subsequent versions of this book. If you find any errata, please report them by visiting `http://www.packtpub.com/submit-errata`, selecting your book, clicking on the **Errata Submission Form** link, and entering the details of your errata. Once your errata are verified, your submission will be accepted and the errata will be uploaded to our website or added to any list of existing errata under the Errata section of that title.

To view the previously submitted errata, go to `https://www.packtpub.com/books/content/support` and enter the name of the book in the search field. The required information will appear under the **Errata** section.

# Piracy

Piracy of copyrighted material on the Internet is an ongoing problem across all media. At Packt, we take the protection of our copyright and licenses very seriously. If you come across any illegal copies of our works in any form on the Internet, please provide us with the location address or website name immediately so that we can pursue a remedy.

Please contact us at `copyright@packtpub.com` with a link to the suspected pirated material.

We appreciate your help in protecting our authors and our ability to bring you valuable content.

# Questions

If you have a problem with any aspect of this book, you can contact us at `questions@packtpub.com`, and we will do our best to address the problem.

# 1
# Getting Started with ArcGIS

All over the world, **Geographic Information Systems (GIS)** are used by small and large organizations alike to manage the environment and to support decision-making in different industrial sectors, such as healthcare, transportation, utilities, communications, petroleum, minerals, and even real estate, banking or insurance. GIS tools are also used by the academic and non-academic institutions in their research projects or disciplines (for example, geology, biology, history, environmental sciences, urbanism, cartography, or cadaster).

Environmental Systems Research Institute (Esri) was founded as a company by Jack Dangermond in the late 1960s. In 1982, Esri released their first commercial software called ArcInfo that had a command-line interface. In the early 1990s, Esri released their first desktop solution with a graphical user interface called ArcView GIS. The ArcView software made the GIS tools more accessible to local administration, academic environments, students, and ordinary users. In 1999, Esri transformed ArcInfo into a modular, scalable desktop and enterprise platform called ArcGIS 8.x. The ArcGIS Desktop 8.x version had three levels of functionality and cost: ArcView, ArcEditor, and ArcInfo. All these levels of functionality shared the same three applications: ArcMap, ArcCatalog, and ArcToolbox.

In 2012, Esri released the ArcGIS 10.1 version. Starting with this version, each ArcGIS for Desktop license includes an ArcGIS Online organizational account with a number of named users and service credits. ArcGIS Online is a cloud-based GIS service that gives organizations the necessary tools to collect real-time data and discover, visualize, create, combine, analyze, manage, and share geospatial information (source: www.esri.com).

In 2015, Esri released the ArcGIS Pro application along with the ArcGIS for Desktop 10.3 version. ArcGIS Pro allows users to work with geospatial data in 2D and 3D environments within the same application.

In the last few years, ArcGIS became a powerful integrated Web GIS platform, which gathers different others technologies, such as **Global Navigation Satellite System (GNSS)**, remote sensing, for example, LiDAR — Light Detection and Ranging, web services, wireless communications, and handheld or mobile devices.

In this chapter, we will cover the fundamental aspects of the **ArcGIS for Desktop** software. We will explore what is needed in terms of hardware and software. We will also show the reader how to install and activate a 60-day trial of ArcGIS for Desktop Advanced Single Use version. By the end of this chapter, you will be ready to run the software and understand the functionality of the main ArcGIS for Desktop applications. We will cover the following topics in this chapter:

- Hardware and software requirements
- Installing ArcGIS for Desktop
- Exploring ArcGIS for Desktop

# Hardware and software requirements

Before you begin installing ArcGIS for Desktop, you need to check whether your computer meets the minimum hardware and software requirements to properly install and run the ArcGIS applications.

# Hardware requirements

In this section, we will list the minimum hardware requirements, and we will check the computer system specifications using the **System Information** tool in Windows 8.

To install ArcGIS for Desktop, we need the following minimum hardware requirements:

- **Central processing unit (CPU) speed**: 2.2 GHz
- **Processor**: x86 or x64 with SSE2 Extensions
- **Memory/RAM**: 2 GB; ArcGIS for Desktop exists only as a 32-bit application and, as such, is limited, only being able to use up to 4 GB of RAM
- **Screen resolution and display properties**: 1024x768 pixels at normal size (96 dpi), 24-bit color depth
- **Swap space**: 500 MB
- **Disk space**: 2.4 GB

- **Video/Graphics Adapter**: 64 MB RAM; ArcGIS for Desktop will work better on a 24-bit capable graphics accelerator with a minimum of 256 MB dedicated video card memory than on integrated graphics

- **Networking Hardware**: TCP/IP Protocol, Network Interface Card (NIC) for the license manager (for Concurrent Use) or authorization information (for Single Use)

Use the **System Information** panel to get information about your system, as shown in the following screenshot:

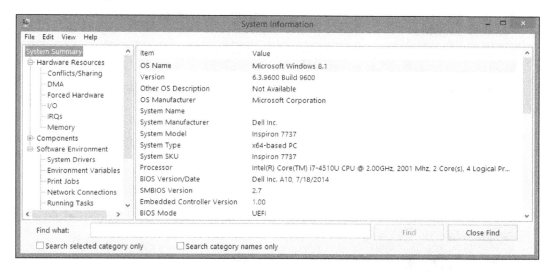

1. To begin, right-click the **Start** button and select **Run**.
2. In the **Run** box, type msinfo32, as shown in the following screenshot:

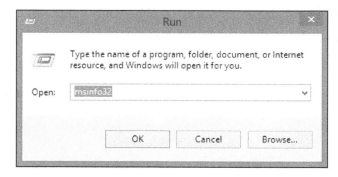

3. Click on **OK** to open the **System Information** panel.

# Software requirements

In this section, we will list the minimum software requirements and the supplementary disk space necessary to install additional components.

To install ArcGIS for Desktop 10.4, we need the following minimum software requirements:

- **Operating system**: 32-bit or 64-bit versions of Windows 7, 8, 8.1, or 10, and Windows Server 2008, and 2012
- **Admin privilege**: Administrative privileges are required to install this software
- **Framework**: Microsoft .NET Framework 4.5 or higher must be installed prior to installing ArcGIS for Desktop
- **Internet Browser**: Esri indication is Internet Explorer 9 or higher; however, Mozilla Firefox, or Google Chrome could be also used

The ArcGIS geoprocessing tools require Python 2.7.10, Numerical Python (NumPy) 1.9.2, and Matplotlib 1.4.3 to be installed.

[  ArcGIS for Desktop installation wizard will automatically install the Python components. ]

# Installing ArcGIS for Desktop

In this section, we will cover the steps necessary to obtain and install a 60-day trial of ArcGIS for Desktop (Advanced Single Use) on Windows.

Esri license authorizes to run ArcGIS for Desktop as the following:

- **Single Use**: This is when software and its extensions are authorized to run on a single computer.
- **Concurrent Use**: This is when a license server manages a given number of floating licenses through a computer network. The computers that have ArcGIS for Desktop installed are authorized by the license server to run the applications (for example, ArcMap) using the ArcGIS License Manager.

# Obtaining a 60-day trial of ArcGIS for Desktop

Follow these steps to obtain a trial of ArcGIS for Desktop:

1. Open your browser and type in the following address: `www.esri.com`.

2. Navigate to **Products | ArcGIS for Desktop**.

3. Click the orange **60-Day Free Trial** button or click the **Free Trial** tab, as shown in the following screenshot:

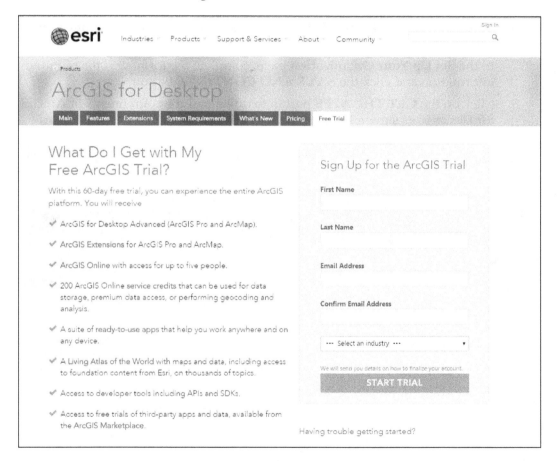

4. Provide the required information and click the **START TRIAL** button.

   You will receive a confirmation e-mail from Esri Customer Service with the subject *Esri – Activate Your Free ArcGIS Trial*. Check your e-mail and use the link to activate your ArcGIS Online account.

5. Choose and type the username and password for your public account. In the **Organization** field, specify the name of your organization (for example, Student_Your Name).You will be the administrator of your own organization.

 Do not select an existing organization from the drop-down list. If you select the name of an organization from the drop-down list, you will need to request permission from the selected organization.

6. Carefully read the **Terms of Use** and check **I accept the terms & conditions**. Click the **CREATE MY ACCOUNT** button.

7. In the **Set Up Your Organization** page, complete the form with the required information and click the **SAVE AND CONTINUE** button.

8. Click on the **GET THE APPS** button to download the components of ArcGIS for Desktop, as shown in the following screenshot:

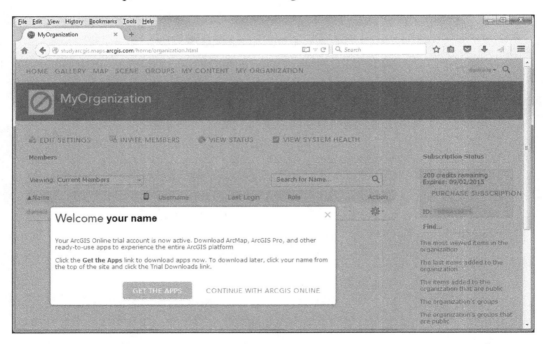

9. We will download the **ArcMap** component, as shown in the following screenshot:

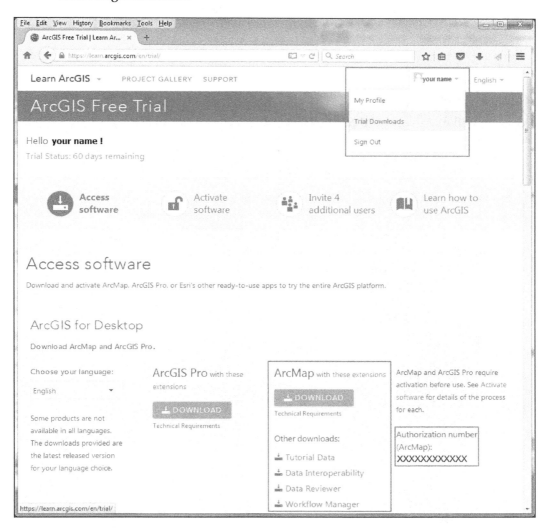

10. Click on the **DOWNLOAD** button and save the archived software to your hard disk. The download process may take a while depending on your internet speed. Please note that you will receive an authorization number for the ArcMap application. We also recommend that you download all the other components, such as **ArcGIS Pro** and **Data Interoperability**, for further study.

If you want to obtain an Esri Technical Certification and need to learn for more than 60 days, we recommend that you purchase a license ArcGIS for Desktop Home from the following link:

`http://www.esri.com/software/arcgis/arcgis-for-home.`

# Installing on Windows

Now, we will install and activate ArcGIS for Desktop Advanced Single Use. Remember that the term "*Advanced*" refers to the license level of the applications that are included with ArcGIS for Desktop: ArcMap, ArcCatalog, ArcGlobe, and ArcScene. Follow these steps to install this software:

1. Double-click the self-extracting `.exe` file named `ArcGIS_ Desktop_103x_ xxxxxx.exe` and click the **Run** button to unpack the installation files.

2. In the second panel, select the destination folder for the extracted files and click **Next**. Keep the **Launch the setup program** option checked and click the **Close** button to start the installation wizard, as shown in the following screenshot:

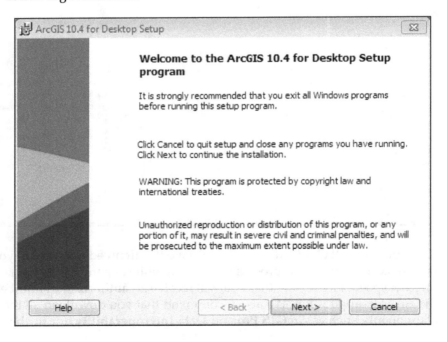

3. Follow the next panels to install ArcGIS for Desktop. The installation may take several minutes.

4. Click on the **Finish** button to exit and launch **ArcGIS Administrator Wizard**, which is shown in the following screenshot:

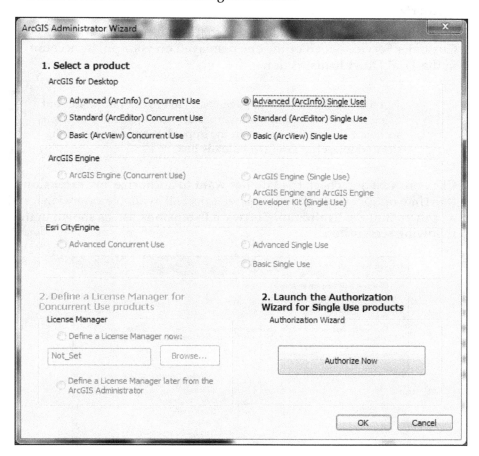

The **ArcGIS Administrator Wizard** window will help you to activate the ArcGIS for Desktop before using it.

If you accidentally closed the **Administrator Wizard** window or you want to authorize the product later on, you can manually start **ArcGIS Administrator** by navigating to **Start | All Programs | ArcGIS | ArcGIS Administrator**. Select **Desktop** and check the **Advanced (ArcInfo) Single Use** option.

5. Click **Authorize Now**. In the **Authorization Options** panel check **I have installed my software and need to authorize it**. Click **Next** and check the **Authorize with Esri now using the Internet** option.

6. Click on **Next** and for the next two panels, complete the forms with the required information. Click on **Next** to see the panel, **Software Authorization Number**.

7. Enter the software authorization code that you received in the Esri Customer Service e-mail or the one displayed on your public account in the **Trial Downloads** section.

 To return to the download page, log in in to your public account using the username and password that you chose in the previous section. Click on your name in the upper left-hand corner of the site and select the **Trial Downloads** link.

8. Click on **Next** and check the **I do not want to authorize any extensions at this time** option. Go to the next panel, select all available extensions, and add them into the **Evaluation Software Extensions** list, as shown in the following screenshot:

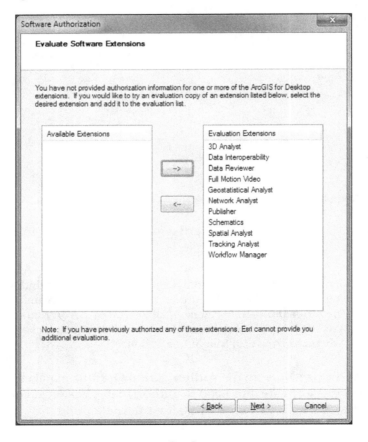

9. Go to the last panel to initiate the authorizing software. Click on **Finish** when the authorizing process is complete. To close the ArcGIS Administrator window click on **OK**.

Now, your ArcGIS for Desktop Advanced Single Use is ready to run.

Depending on your ArcGIS for Desktop version, please download and install the latest patches (bug fixes) or service packs (compilation of bug fixes) from `http://support.esri.com/en/downloads/patches-servicepacks`.

# Exploring ArcGIS for Desktop

As you saw in **ArcGIS Administrator Wizard**, ArcGIS for Desktop has three levels of functionality: Basic (formerly ArcView), Standard (formerly ArcEditor), and Advanced (formerly ArcInfo). All these three products comprise two main applications: **ArcMap** and **ArcCatalog**. These two applications look and work the same for the three license levels but are differentiated by the functionality that they provide. The 10.3 version of ArcGIS for Desktop released a third new application called **ArcGIS Pro** in January 2015.

This book will not cover the ArcGIS Pro concepts.

If you want to learn about ArcGIS Pro, we recommend the book *Learning ArcGIS Pro, Tripp Corbin, Packt Publishing*, which can be found at `https://www.packtpub.com/application-development/learning-arcgis-pro`.

# ArcCatalog

The ArcCatalog application allows users to explore, manage, and even document their spatial or nonspatial data. The interface of ArcCatalog is shown in the following screenshot:

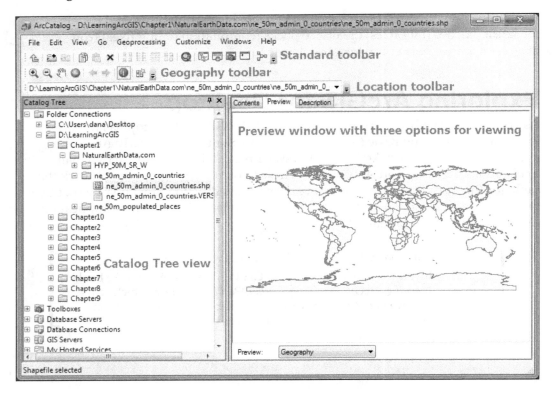

**Catalog Tree** shows the folders and data files, geodatabases, database connections, GIS servers, and custom and system toolboxes. Think of the Catalog Tree as the Windows Explorer of ArcGIS. The Catalog Tree window can be accessed in the ArcMap application, too.

The **Preview** window allows you to view and explore the geometric shapes and attributes of features stored in the dataset selected in the Catalog Tree.

ArcCatalog has three options to view the selected items in the **Preview** window:

- **Contents**: This is a quick view of the content of a selected item in the Catalog Tree (for example, a folder or geodatabase content)

- **Preview**: This is used to view and explore the geography (geometry) and attributes of features stored in the dataset selected in Catalog Tree

- **Description**: This is used to view and edit the information about data, known as **metadata** or "data about data"

> This book includes datasets that you will need to complete the exercises here. Before you can work with these datasets, you have to download them from your account at http://www.packtpub.com.
>
> If you purchased this book elsewhere, then you can visit https://www.packtpub.com/books/content/support and register there to have these files e-mailed directly to you.
>
> This training dataset will require at least 10 GB of free space on your working hard disk. Extract the training data from the archive on your working disk in a new folder named LearningArcGIS.

Follow these steps to add some publicly available data from NaturalEarthData.com:

1. From www.naturalearthdata.com/downloads/, download the following themes after navigating to **Medium scale data (1:50m)** | **Cultural**—Admin 0-Countries (click the **Download countries** button) and Populated Places (click the **Download populated places** button). Also, download the Natural Earth 1 with Shaded Relief raster from the **Raster\Natural Earth 1** section.

2. Unzip the archive files to <drive>:\LearningArcGIS\Chapter1\ NaturalEarthData.

> If you didn't succeed in downloading and unzipping the data, you can find the necessary datasets at <drive>:\LearningArcGIS\ Chapter1\Results.

Follow these steps to explore the data in the ArcCatalog application:

1. Start the ArcCatalog application by navigating to **Start** | **All Programs** | **ArcGIS**.

2. To see the downloaded datasets in the panel from the left-hand side of the window named **Catalog Tree**, click **Connect To Folder** from the **Standard** toolbar. Go to <drive>:\LearningArcGIS\Chapter1 and click **OK**. In **Catalog Tree**, expand the NaturalEarthData\ ne_50m_admin_0_countries folder and select the polygon shapefile named ne_50m_admin_0_countries. shp, as shown in the previous screenshot.

3. If you don't see the file extension of shapefile (`*.shp`), go to the **Customize** menu, select **ArcCatalog Options**, click the **General** tab, and uncheck the option, **Hide file extension**. Click **OK** to close **ArcCatalog Options**.

 The shapefile format (`.shp`) is an Esri vector data format. Even if ArcCatalog displays it as a single file, shapefile is a set of differently-related files that store geometric and attribute information of the geographic features. For more information about the shapefile format, please refer to *ESRI Shapefile Technical Description, ESRI White Paper-July 1998* at `https://www.esri.com/library/whitepapers/pdfs/shapefile.pdf`.

4. In the right-hand side of the window, select the **Preview** tab to see the map of the countries. Use the **Zoom In**, **Zoom Out**, and **Pan** tools from the **Geography** toolbar to explore data in the preview window.

5. From the **Geography** toolbar, click the **Identify** tool. Then, click on any country in the preview window. The **Identify Results** window shows the attributes of the country that you have just clicked.

6. If you want to see the attributes for all countries, click the drop-down arrow next to **Preview** from the bottom of the preview window and select **Table**. You can now see the attribute table of all 241 countries.

7. Let's examine these attributes. Scroll through the table and select the `name` field. We will search the `United Kingdom` value in the selected field. Click the **Table Options** button from the bottom left corner of the preview window and select **Find**. In the **Find** window, for **Find what**, type `United Kingdom`. Check the **Search Only Selected Field(s)** option and click **Find Next** to search for the record. The record with the `FID=77` value and the `name = United Kingdom` field is underlined in the attribute table. When finished, click **Cancel** to close the **Find** window.

8. Return to the geography view mode by selecting **Geography** from the drop-down arrow next to **Preview**.

9. Repeat these steps to explore the shapefile named `ne_50m_populated_places.shp` and the shaded relief file named `HYP_50M_SR_W.tif`.

 The shapefile's green icon next to the shapefile's name indicates what type of geometry is stored in it: point, polyline, or polygon. Also, the color of the icons is tied to the data format. For example, the geodatabase icon is gray, and the CAD files are blue.

10. Close the ArcCatalog application.

# ArcMap

The ArcMap application allows users to visualize, create, edit, query, and analyze spatial or nonspatial data. ArcMap works with map document files. A **map document** file (.mxd) stores the reference to the data source (for example, the reference to the ne_50m_admin_0_countries.shp shapefile), the reference to your symbol styles, how your data is displayed and symbolized on a map, and other properties related to your data.

 A map document does not store a copy of your data.

The ArcMap interface is shown in the following screenshot:

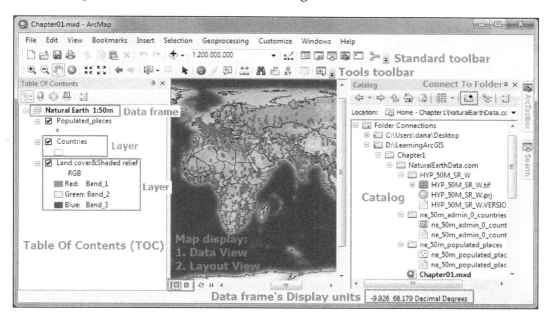

In ArcMap, a **Data frame** is a container for the layers displayed on a map. Think of the data frame as a classic map. All layers from a data frame will be displayed in the same coordinate system, which is the property of the data frame. In one map document, you can have multiple data frames with different coordinate systems.

A **Layer** is a visual display of your dataset in ArcMap. For example, you can add the same shapefile, named `ne_50m_admin_0_countries.shp`, three times in ArcMap, and the **Table Of Contents** window will automatically display them as three layers, symbolized with different symbols, as shown in the following screenshot:

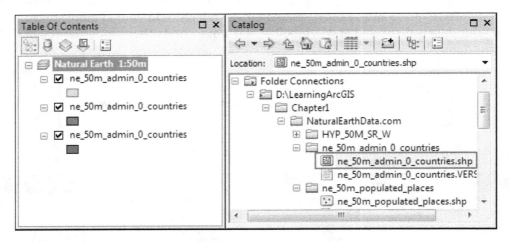

A layer can be stored in a **map document** (`*.mxd`) or as an individual **layer file** (`*.lyr`). ArcMap has two methods of map display: **Data View** and **Layout View**. In the **Data View** mode, you can view, create, edit, symbolize, and analyze data from only one data frame at a time. In the **Layout View** mode, you can use multiple data frames to create final professional maps (map layout) for print or digital view.

 Note that all steps in this subsection have to be completed in one session.

Follow these steps to add spatial data as layers to ArcMap:

1. Start the ArcMap application by navigating to **Start | All Programs | ArcGIS** or using the **ArcMap** button on the **Standard** toolbar, in ArcCatalog application.

2. In the **ArcMap-Getting Started** window, select the **Blank Map** user template from **New Maps | My Templates**, as shown in the following screenshot:

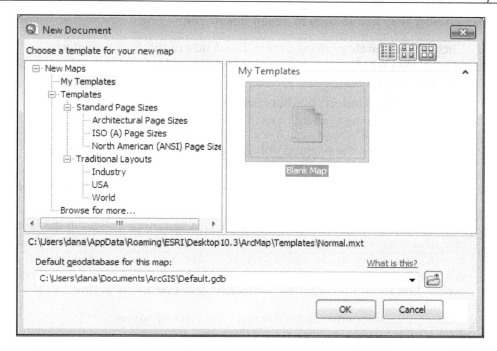

3.  Click on **OK**. The ArcMap interface will open with an empty and untitled map document.

4.  Click on the **Add Data** button from **Standard** toolbar. Double-click **Folder Connections** and select the HYP_50M_SR_W.tif raster from <drive>:\ LearningArcGIS\Chapter1\NaturalEarthData\HYP_50M_SR_W. Click the **Add** button.

5.  The land cover raster has been added in the **Table of Contents** window as a layer in the default data frame named Layers. To change the name of the data frame, click the word Layers to select it, then click it again, and type Natural Earth 1:50m.

If you accidentally double-click the data frame's name, the **Data Frame Properties** window will be opened. If you double-click the layer's name, the **Layer Properties** window will be opened.

6.  Change the name of layer from HYP_50M_SR_W.tif to Land cover&Shadedrelief by following the previous step.

Let's try another way to add data to a map document:

1. Click on the **Catalog** tab on the left-hand side of the ArcMap interface. The **Catalog** window can be also opened using the **Catalog** button from the **Standard** toolbar.

2. In the **Catalog** window, double-click **Connect to Folder** and select the ne_50m_admin_0_countries.shp shapefile from <drive>:\ LearningArcGIS\Chapter1\NaturalEarthData\ne_50m_admin_0_ countries. Drag the selected polygon shapefile and drop it into the map display area.

3. Repeat the previous step to add ne_50m_populated_places.shp shapefile to **Table Of Contents** as a layer.

4. You can reordering map layers in **Table Of Contents** by clicking the layer and dragging it below or above the existing layers. You can order the map layers, only if they are in **List by Draw Order**. You cannot adjust the layer order in **Table of Contents** if they are listed in the following ways of listing layers: **List By Source**, **List By Visibility**, and **List By Selection**.

We will now change the name and symbols of the last two layers:

1. In **Table of Contents**, right-click the ne_50m_admin_0_countries.shp layer and select **Properties**.

2. In the **Layer Properties** window, click the **General** tab; and for **Layer Name**, replace the existing name by typing Countries. Changing the layer name does not change the name of the source dataset. Click **Apply**.

3. Click on the **Symbology** tab, and for **Symbol**, click the **rectangular symbol** button to open the **Symbol Selector** window, as shown in the following screenshot:

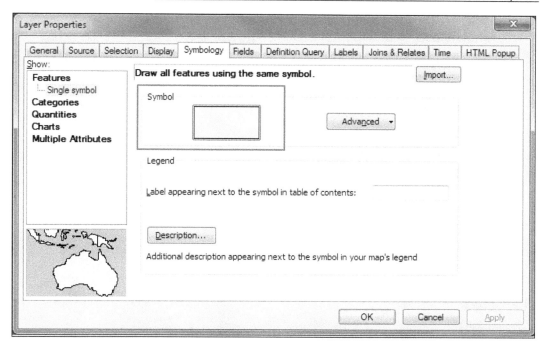

4.  In the **Symbol Selector** window, select the **Olive** style and click **OK** to close the **Symbol Selector** window.

You may have noticed that the Countries layer is obscuring the shaded relief layer in the map display. You will now add transparency to the Countries layer symbols:

1.  In the **Layer Properties** window, click the **Display** tab; and for **Transparent**, type 50. Click the **Apply** button to apply the changes to the layer and click **OK** to save the applied changes and close the **Layer Properties** window.

2.  Use the **Zoom In**, **Zoom Out**, and **Pan** tools from the **Tools** toolbar to inspect the results in the map display area.

3. Click on the **Find** tool from the **Tools** toolbar and set the parameters, as shown in the following screenshot:

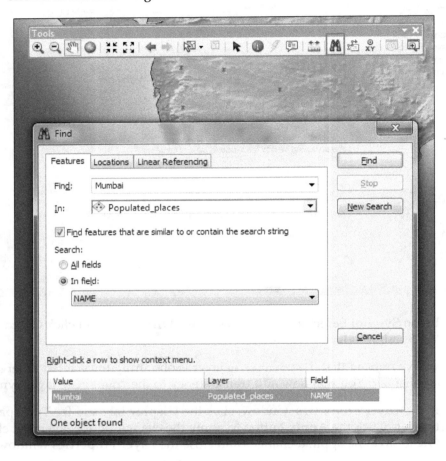

4. Check whether the **Features** panel is active; and for the **Find** text box, type Mumbai. For the **In** drop-down list, select the Populated_places layer (ne_50m_populated_places.shp), and for the **In field** drop-down list, select **NAME**. Click **Find**. At the bottom of the **Find** window, right-click the displayed result and select **Zoom To**. The city has been centered in the map display.

5. If you want to see the attributes of the point feature, right-click the search result and select **Identify**. In the **Identify** window, scroll through the fields list and search for the **ADMoNAME** field to see the name of the country. Click **Cancel** to close the **Find** window.

6.  Let's confirm the country name using the information stored in the
    Countries layer. Click the **Identify** tool and click the map display.
    For **Identify**, from the drop-down list, select the **Countries** layer
    (ne_50m_admin_0_countries.shp).

Now let's explore another part of the world using their geographic coordinates,
as shown in the following screenshot:

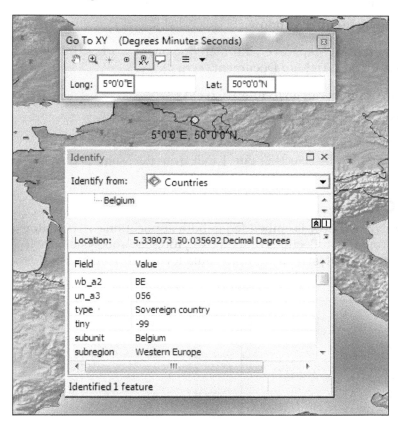

The coordinate system of our datasets is **WGS 1984**, which is a geographic coordinate
system, and units of measurement are expressed in degrees. Let's search for a location
using geographic coordinates:

1.  Select **Go To XY** from the Tools toolbar. By default, the geographic
    coordinates are expressed as **Degrees Minutes Seconds**. You can change
    this to **Decimal Degrees** by clicking the drop-down arrow from the top of
    the **Go To XY** window. Under the field **Long:**, type 5 degrees, and for **Lat:**,
    type 50 degrees. Press the *Enter* key.

2. Click the **Add Labeled Point** button under **Go To XY** window to add a graphic point with information about its geographic coordinates in the map display.

3. To get some additional information about the area around the point, select the **Identify** tool and click the **green graphic point**. For **Identify** window, from the drop-down list, select the Countries layer. What is the name of the country that you selected?

4. Close the **Identify** window.

Before closing the ArcMap, save your map document work, symbols, and green point labeled into an ArcMap Document (*.mxd):

1. From Main menu, go to **File | Save As...**.

2. Go to <drive>:\LearningArcGIS\Chapter1\NaturalEarthData.

3. Name the file as MyChapter01.

4. Exit ArcMap by selecting **File | Exit** from the Main menu.

# ArcToolbox

The ArcToolbox window is embedded in both applications: ArcCatalog and ArcMap. ArcToolbox contains collections of toolboxes, toolsets, and system tools for data management and analysis, cartography, data format conversion, and other tasks, as shown in the following screenshot:

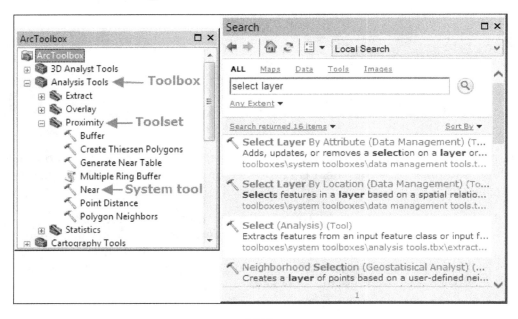

To increase its efficiency, ArcToolbox should be used in conjunction with the **Search** and **Results** windows. Both are available in ArcCatalog and ArcMap applications, from the **Geoprocessing** menu.

Some tools require their corresponding extensions to be enabled. For example, if you want to use the **3D Analyst** processing tools, you should manually enable the **3D Analyst** extension before working with them. To enable this extension in the ArcCatalog application, click the **Customize** menu and select **Extensions**. In the **Extensions** window check the **3D Analyst** extension and click **Close**. The same step is also available in the ArcMap application.

# Getting help

Getting help while you are working with ArcGIS for Desktop applications will increase productivity. If you are stuck or you want to quickly get more information about a tool or a button, then use *ArcGIS Help* which is accessible from all ArcGIS for Desktop applications.

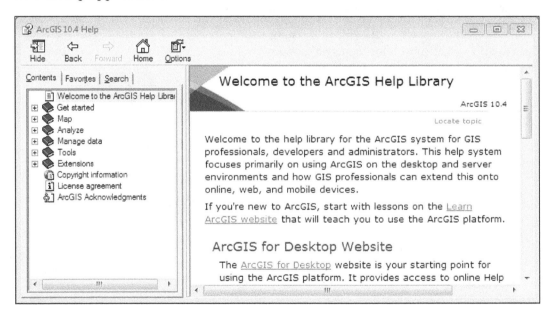

There are three tabs in the upper right-hand corner of the **ArcGIS Help** window:

- **Contents**: This is where the help topics are organized into books on the left-hand side panel
- **Favorites**: This stores a list of your favorite or most-viewed help topics

- **Search**: This searches for a help topic by entering one or more keywords

 To discover the ArcGIS community (for example, Web help, blogs, and forums), use ArcGIS Resource Center, which is available at http://resources.arcgis.com.

# Summary

In this chapter, you learned how to obtain, install, and activate a 60-day trial of ArcGIS for Desktop Advanced Single Use version.

After installing this, you explored some publicly available data from NaturalEarth Data using the main ArcGIS for Desktop applications: ArcCatalog and ArcMap. In ArcCatalog, you learned how to explore the geometry and attributes of your spatial data. In ArcMap, you learned two different ways to add spatial data as layers to ArcMap. You used the **Identify** tool in conjunction with **Find** and **Go To XY** tools to access information about two different countries.

In the next chapter, we will explore the geographic principles, such as geodetic datum and coordinate reference systems, and we will learn to work with different map projections.

# 2
# Using Geographic Principles

This chapter will provide you with an introduction of the geographic principles, such as coordinate system, spheroid, coordinate reference system, geodetic datum, geoid, and map projections. We will learn to distinguish between geographic and projected coordinate systems, and we will use the datum transformations to correctly convert and transform different coordinate reference systems (CRS). We will also apply different map projections to the same dataset to visually analysis the patters of distortion.

By the end of this chapter, you will learn the following topics:

- Using coordinate systems
- Geodetic datum and geoid
- Using coordinate reference systems
- Using map projections
- Transforming coordinate reference systems

## Using coordinate systems

A **coordinate system** (CS) defines a set of axes that span the coordinate space (Iliffe, 2003). A CS defines the attribute of axes, such as number of axes (dimensions: 1D, 2D, or 3D), direction, names, units, and even the order of axes. The most common systems used in GIS are as follows (Allan A.L., 2007):

- Cartesian coordinates (X, Y, Z)
- Geodesic coordinates (geodetic latitude and longitude)
- Spherical coordinates (geocentric latitude and longitude)

- A hybrid 2D coordinate system and 1D coordinate system on a map projection (for example, E=Easting, N=Northing, and H= orthometric height)
- A hybrid 2D coordinate system and 1D coordinate system for the ellipsoid and sphere (for example: φ, λ, and h= ellipsoidal height)

 For theory aspects regarding coordinate systems, datum, and map projections, please refer to: Jonathan Iliffe and Roger Lott; *Datums and Map projections for remote sensing, GIS, and surveying 1st Edition*; Whittles Publishing, 2003.

Mathematic cartography uses the **ellipsoid** of revolution (**spheroid**), **sphere**, and plane as "reference surfaces" to define coordinate systems.

# The ellipsoidal coordinate system

The first approximation of the shape of the Earth is that it is a rotational ellipsoid or a reference ellipsoid. The following figure shows the **graticule** of parallels and meridians at $10^0$ intervals:

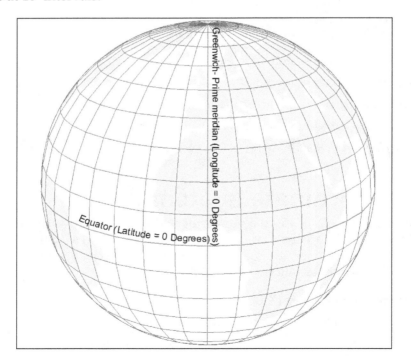

The surface of revolution is obtained by rotating the **meridian ellipse** around its minor (short) axis. This ellipsoid is called an **oblate ellipsoid**. The major and minor axes of a reference ellipsoid do not vary greatly. This is because its shape approximates the shape of a sphere, so the terms "ellipsoid" and "spheroid" are often used interchangeably by Esri's software.

The size and shape of the ellipsoid can be defined by at least two geometric parameters, where one should be a semi-axis. For example, the most common ellipsoid, called **GRS 1980,** has the following parameters: its semi-major axis is 6,378,137 m and its flattening is 1/298.25722 21008 827 (Hooijberg, M. Practical Geodesy, 1997).

The ellipsoidal coordinate system or geodetic coordinate system can be two-dimensional (**geodetic latitude,** $\phi$ or $\varphi$ and **geodetic longitude,** $\lambda$) or three-dimensional ($\phi$, $\lambda$, and **ellipsoidal height** h), as shown in the following figure:

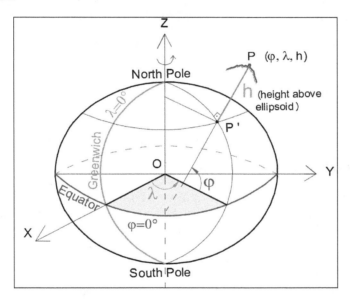

The latitude and longitude are expressed as sexagesimal degrees in minutes and seconds (DMS), or they are expressed as sexagesimal **decimal degrees** (DD), especially for the digital storage and computation, as shown in the following screenshot:

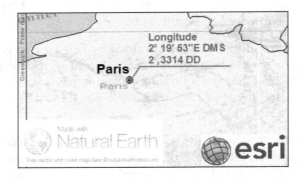

This example shows the geographic coordinate for Paris in DMS and DD. A short example of converting from 2°19'53" E DMS to DD is explained, as follows:

1. Divide the minute value by the number of minutes in a degree (60):

   *19 minutes= 19/60 = 0.316666 degrees*

2. Divide the seconds value by the number of seconds in a degree (3600):

   *53 seconds=53/3600=0.014722 degrees*

3. Add up the degrees: $2^0 + 0.31\ 66\ 66^0 + 0.01\ 47\ 22^0 = 2.33\ 13\ 88\ DD \sim 2.3314^0\ DD$

The ellipsoidal height is expressed in meters.

The Equatorial plane splits the ellipsoid into two hemispheres: the Northern hemisphere (N) where latitude has values between $[0^0, +90^0]$, and the Southern hemisphere (S) where latitude has values between $[0^0, -90^0]$.

The Greenwich meridian plane (**prime meridian**) splits the ellipsoid into two hemispheres: the Eastern hemisphere (E. Greenwich) where longitude has values between $[0^0, +180^0]$, and the Western hemisphere (W. Greenwich) where latitude has values between $[0^0, -180^0]$.

# The spherical coordinate system

A more simplified approximation of the shape of the Earth is the sphere. The spherical coordinate system can be two-dimensional (latitude φ and longitude λ) or three-dimensional (latitude φ, longitude λ, and height above sphere, h), as shown in the following figure:

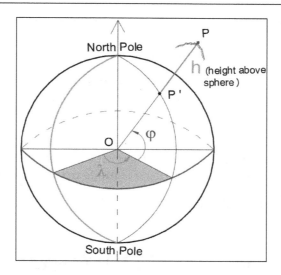

A sphere has a constant curve on any latitude, in every direction, which allows generalizations in map-projection formulae. In the ArcMap coordinate system list, you will find the term "**authalic**", which is used for some coordinate reference systems. An authalic sphere is a sphere with a surface area equal to the surface area of the given ellipsoid. For example, the **GRS 1980 Authalic Sphere** has the following parameters: the semi-major and semi-minor axes are 6,371,007 m and the flattening is 0.

# The spherical polar coordinate system

The spherical polar coordinate system is an oblique or transverse coordinate system that is defined by a chosen pole on a sphere called Q that has the latitude on the sphere, $0^0 \leq \varphi < 90^0$. The spherical polar coordinates are **azimuth** (A) and **zenith** distance (Z), as shown in the following figure:

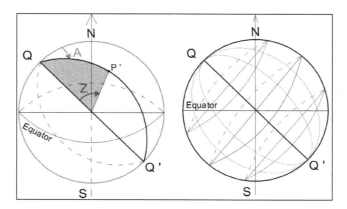

Azimuth (A) is the angle from the arc of a great circle that crosses through the pole Q, point P (geodesic line QP'), and the meridian of pole Q. The angle is measured clockwise and has values between $[0^0, 360^0]$.

Zenith distance (Z) is the central angle that subtends the arc of the great circle from pole Q to a given point on sphere P'. The subtended angle Z has values between $[0^0, 180^0]$.

**Verticals** (A is constant) are arcs of a great circle between the poles Q and Q'. **Almucantars** (Z is constant) are small circles with variable radius, and they are perpendicular to the verticals. The graticule is made up of verticals and almucantars.

The spherical-polar coordinate system is an additional coordinate system that is used by the traverse and oblique map projections.

# The three-dimensional (3D) Cartesian coordinate system

The 3D Cartesian coordinates (X, Y, Z) rotate and are attached to the model of the Earth. The origin of coordinate system is the center of the ellipsoid or sphere. The plane XOY is in the Equatorial plane. The OX-axis is a line from the center that runs through the intersection of the Greenwich meridian with Equatorial plane. The OY-axis is perpendicular to the OX-axis. The OZ-axis coincides with the minor axis of the ellipsoid (polar axis), as you can see in the following figure:

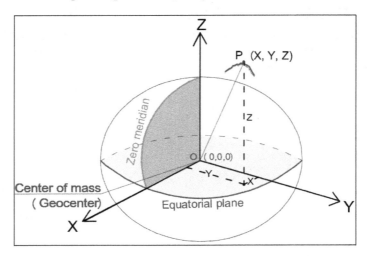

In most of these cases, the 3D Cartesian coordinate system coincides with the center of mass of the Earth called **geocenter**. In this case, the OX-axis is toward $90^0$ E longitude, and the OZ-axis is a line from the geocenter through the **Conventional Terrestrial Pole** (which coincides with the Earth's axis of rotation). This Cartesian coordinate system is known as **Terrestrial Reference System (TRS)**. The Cartesian coordinates are expressed in meters.

# Geodetic datum and geoid

The international standard ISO 19111:2007 called Geographic Information: Spatial Referencing by Coordinates defines **datum as**: "a set of parameters that defines the position of the origin, the scale, and the orientation of a coordinate system."

# Global and regional datums

A datum can be a **horizontal datum** or a **vertical datum**, either global or local. Jekeli (2006) showed in his previous work that a traditional datum is defined by eight parameters: three to define its origin (for example, the center of mass of the Earth), three to define its orientation, and two to define the ellipsoid.

As you could see, an ellipsoid is not enough to define a global datum. We need an ellipsoid and a Terrestrial Reference System that has the defined origin and orientation with respect to the Earth, as in the following figure:

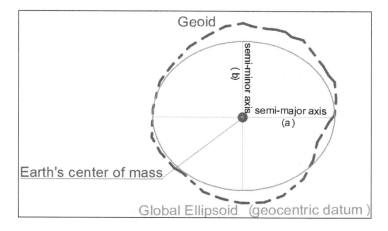

An example of global datum is the **World Geodetic System 1984**, known as **WGS84**.

A regional datum aligns the ellipsoid to a regional surface in such way to have a better match with the surface of the Earth. In this case, the origin of the datum is defined by a physical point (fundamental point) with horizontal geodetic coordinates. For example, a geodetic regional datum called Pulkovo 1942 is defined by the following:

- The latitude and longitude of an initial point (fundamental point): the Pulkovo Observatory, Latitude 59°46'18.55"N and longitude 30°19'42.09"E of Greenwich.

- The azimuth of a line from the initial point to Bugrõ is $\alpha_0$ = 12006'42.305" (Source: Grids&Datums. Republic of Estonia, Clifford J. Mugnier, 2007).

- Two parameters of the reference ellipsoid: Krassowsky 1940, a semi-major axis 6,378,245 meters and an inverse flattening of 298.3.

> There are local datums that use the same ellipsoid but have different points of origin.

Another example of regional horizontal datum is the North American Datum 1927 (NAD27), which is based on the Clarke ellipsoid of 1866. (Source: North American Horizontal Datums by Jan Van Sickle.)

So you have seen how a local or global datum can be used to help determine the X and Y coordinates of features. However, how do we determine the Z or elevation? The *ISO 19111: 2007 Geographic Information Spatial Referencing by Coordinates* standard defines a vertical datum as the relation of the gravity-related heights to the Earth and is related to a geoid. So, a vertical coordinate reference system (1D) is then based on a vertical datum in the same way that a horizontal one is.

# The geoid and heights

The international standard ISO 19111:2007 called Geographic Information Spatial Referencing by Coordinates defines a geoid as: "the equipotential surface of the Earth's gravity field that is everywhere perpendicular to the direction of gravity. The geoid best fits mean sea level either locally or globally. A geoid is considering the true shape of the Earth."

> For a quick introduction of the mean sea level term, please refer to the following article: *Mean Sea Level, GPS, and the Geoid* by Witold Fraczek, Esri Applications Prototype Lab, at http://www.esri.com/news/ arcuser/0703/geoid1of3.html.

The relationship between the ellipsoid, geoid, and topographic surface in terms of height, is described in the following figure:

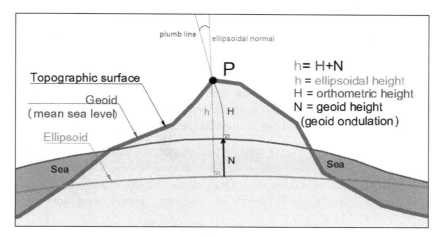

# Using a coordinate reference system

A **coordinate reference system (CRS)** is defined by the European Directive 2007/2/EC as "a system for uniquely referencing spatial information in space as a set of coordinates (X, Y, Z) and/or latitude, longitude, and height, based on a geodetic horizontal and vertical datum".

> For a short introduction of coordinate reference systems (CRS definition, conversion and transformation), please refer to the following URL:
>
> http://www.crs-geo.eu/nn_124210/crseu/EN/CRS__
> Overview/crs-overview__node.html?__nnn=true

In this section, we will explore five coordinate reference systems (CRS) that are based on different horizontal datum. The terms coordinate reference systems and geographic coordinate systems are used interchangeably by the Esri's software.

Follow these steps to start exploring the datum and coordinates reference system with the ArcMap application:

1. Start the ArcMap application to open an existing map document.
   In the **ArcMap -Getting Started** dialog, select **Browse for more** and
   select WorkingWithDatums.mxd from <drive>:\LearningArcGIS\
   Chapter2\Datums.

2. The **Table Of Contents** section contains a single data frame called **Datums** with three layers. The `WorldGrids_10` layer shows the graticule of parallels and meridians at 100 intervals. The `ReferenceLines` layer displays the reference lines: mean equator, and zero meridian. The `Populated_places(Natural Earth 1:50m)` layer contains the major cities and towns in the world at the scale: 1:50,000,000.

3. In **Table Of Contents,** right-click the **Datums** data frame, select **Properties,** and click the **Coordinate System** tab. You can see the parameters of the current coordinate reference system, which is a geographic coordinate system called **WGS_1984**, as shown in the following screenshot:

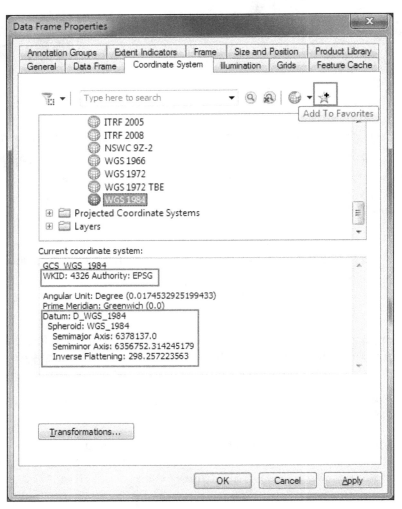

To read more about the **D_WGS_1984** datum, please check the website, http://www.epsg-registry.org. Use the well-known ID of the coordinate reference system WKID: **4326**.

4. At the bottom of coordinate system list, you can see a folder called Layers. This folder contains the list of the coordinate systems used by the layers in the data frame. All three layers share the same **Geographic Coordinate System** as the data frame.

5. You can bookmark your most-used coordinate reference systems by adding them in the Favorites folder. Use the **Add To Favorites** tool by right-clicking **WGS 1984** or by selecting the yellow star. To delete a coordinate reference system from the Favorites folder, right-click it and select **Remove From Favorites**.

6. Click the **General** tab. As you may notice from the status bar, the display units are **Decimal Degrees**, which are determined by the Data Frame's coordinate reference system (**Units Map: Decimal Degrees**). The ArcMap can display the coordinate's values, independent of the original map units. Change the display units to **Degrees Minutes Seconds** by clicking on the drop-down list after navigating to **Units | Display**. Click on **Apply** and then **OK** to close the **Data Frame Properties** window.

7. From the **Customize** menu, select **ArcMap Options** and click the **Data View** tab. For **Round coordinates to** in **Coordinate Display In Status Bar**, choose 3 decimal places. Click on **OK**.

Now, let's add a little bit of color in the background from ArcGIS Online:

8. You need an Internet connection for this step. To check whether ArcGIS for Desktop is connected to ArcGIS Online, right-click **ArcGIS Connection Utility** from the Windows system tray and select **Test Connection Now** to run the connection, as shown in the following screenshot:

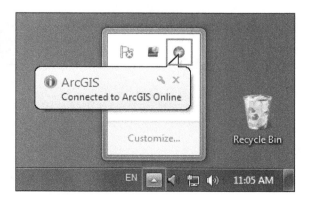

9. On the **Standard** toolbar, from the drop-down arrow next to **Add Data**, select **Add Data from ArcGIS Online**. Search for the Ocean data, select **Ocean Basemap (Mature Support)**, and click on **Add**. Inspect the data using the **Zoom In**, **Zoom Out**, **Pan**, and **Full Extent** tools that are located on the **Tools** toolbar.

We will now customize the list of scales that are displayed in the drop-down list, as can be seen in the following screenshot:

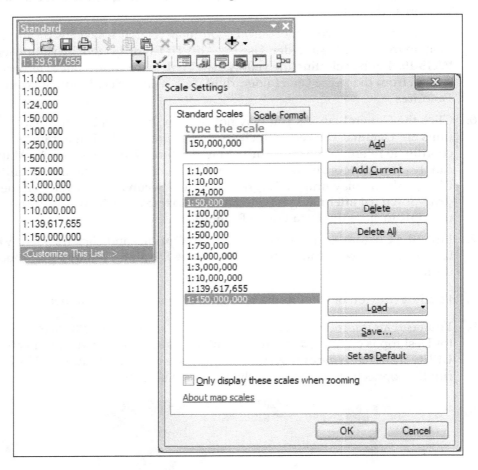

10. You may already know that the scale is the fraction between the distance on a map and the true distance on the Earth. The scale is given as a fraction or a ratio: 1 millimeter [on the map] / 50,000 millimeters [on the ground] or 1:50,000. If we change the unit of measurement on both sides of the fraction, the scale remains same: 1 inch / 50,000 inches or 1: 50,000.

11. From the **Standard** toolbar, click the drop-down list of map scale and select **<Customize This List...>**. Type 50,000 and click the **Add** button to add the value to the map scale list. Again, add the 150,000,000 scale to the list. Click on **OK** to close the **Scale Settings** window.

12. As the Ocean Basemap layer does not provide enough details for a large-scale map display such as 1: 25,000, we will set a visible scale range for it. Select the **1:150,000,000** scale from the map scale list.

13. In **Table Of Contents**, right-click the **Ocean Basemap** layer and navigate to **Visible Scale Range | Set Minimum Scale**. Again, select the **1:50,000** value from the map scale list. Right-click the Ocean Basemap layer and navigate to **Visible Scale Range | Set Maximum Scale**. The layer will be drawn starting from the small-scale 150,000,000 to the large-scale 50,000.

14. For **Scale**, type 150,000,000 and press the *Enter* key. Inspect the results. If you set the map scale at 1:150,000,001 and 1:49,999, then the Ocean Basemap layer will not display and the check box from the left-hand side of the layer is dimmed in **Table Of Contents**.

15. If you want to see the layer at all scales, you have two ways:

    ° Right-click the **Ocean Basemap** layer and navigate to **Visible Scale Range | Clear Scale Range**.

    ° Right-click the layer and select **Properties**. In the **General** section, check the option, **Show layer at all scales**, and click on **OK**.

You can choose the display format of the scales: absolute and relative. The absolute scale is given as 1:6,000. The relative scale is given as 1 inch equals 500 feet. You can set the display format from the drop-down list of map scale by navigating to **<Customize This List...>** **| Scale format | Edit**. ArcGIS automatically converts the relative scale to an absolute scale.

We will now learn how to add the latitude and longitude (geographic coordinates) expressed as sexagesimal degrees, minutes, and seconds (DMS) in the attribute table of the `Populated_places (Natural Earth 1:50m)` layer, as shown in the following screenshot:

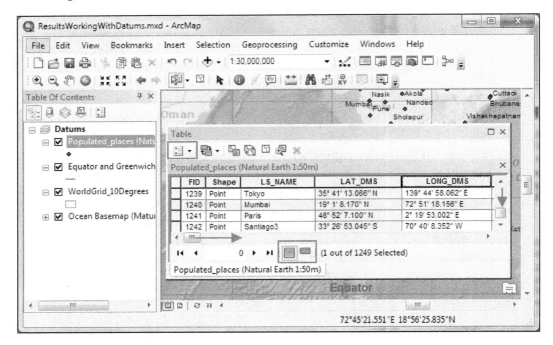

Follow these steps to add the latitude and longitude values for the `Populated_places (Natural Earth 1:50m)` layer:

1. In **Table Of Contents**, right-click the **Populated_places** (`Natural Earth 1:50m`) layer and click **Properties**. Select the **Source** tab. You will find out that the `ne_50m_populated_places.shp` shapefile is the data source of the layer. The coordinate system of the data source is the same geographic coordinate system called **GCS_WGS_1984**. Click on **OK** to close the **Layer Properties** window.

2. Right-click the **Populated_places** layer again and click **Open Attribute Table**. Examine the attribute fields of the layer.

3. All feature geographic coordinates are stored by ArcGIS in decimal degrees. In this step, we would like to store the geographic coordinate as Degrees Minutes Seconds (DMS) in an attribute table using two new fields. Click the **Table Options** button and select **Add Field**. For **Name**, type LAT_DMS; for **Type**, select **Text** from the drop-down list. Click on **OK**. Use the horizontal scroll bar to see the field (last column). Right-click the heading of the **LAT_DMS** field and click **Calculate Geometry**. Click on **OK** to calculate the attributes outside of the edit session and check the **Don't warn me again** option. In the **Calculate Geometry** dialog, for **Property**, select **Y Coordinate of Point**. For **Units**, select **Degrees Minutes Seconds (DDD MM'SS.sss"[N | S])**. Uncheck the **Calculate selected records only** option. Click on **OK**.

4. Repeat the last step to create the LONG_DMS field that stores the longitude using **Property: X Coordinate of Point** and **Degrees Minutes Seconds (DDD MM'SS.sss"[W | E])**. If you want to change the order of the fields in this table, click the heading of the column and drag it to where you want it to be placed.

5. Move and resize the **Table** window in the ArcMap window so that you can see the map display area. From the **Tools** toolbar, use **Select Features** to select a city from the map display area. At the bottom of the **Table** window, click **Show selected records** to view only the selected city.

6. To see all records from the attribute table again, click the **Show all records** button. To unselect the city use the **Clear Selection** button at the top of the **Table** window from the **Table Options** menu or use the **Clear Selected Features** tool, which is next to the **Select Features** tool. Leave the **Table** window open.

In the next steps, you will add three shapefiles with different geographic coordinate systems than the data frame and one shapefile with an unknown coordinate system without choosing a suitable datum transformation. By ignoring the datum transformation, we want to emphasize the horizontal datum shifts and how the geographic coordinates of a given point are different depending upon the geodetic datum, as shown in the following general figure:

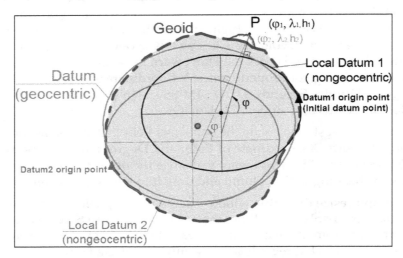

7. From the **Standard** toolbar, click the **Catalog** button. In the **Catalog** window, go to <drive>:\LearningArcGIS\Chapter2\Datum. Right-click **ne_50m_populated_places_GRS80.shp** and select **Properties**. Click the **XY Coordinate System** tab to check the associated coordinate system. The geographic coordinate system is **GCS_ETRS_1989** with the **D_ETRS_1989** datum that is based on the spheroid called **GRS_1980**. Click on **OK** to close the **Shapefile Properties** window.

8. Click and drag **ne_50m_populated_places_GRS80.shp** into the map display. **Geographic Coordinate Systems Warning** displays, as shown in the following screenshot:

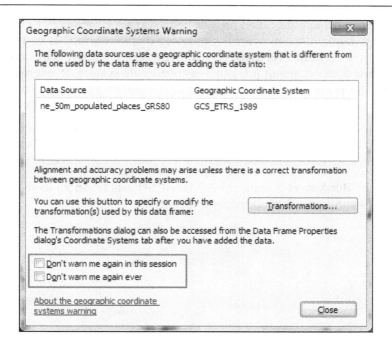

9. ArcGIS needs to change the layer's datum through a datum transformation. If you close **Geographic Coordinate Systems Warning,** no transformation will be applied and the data will be a simple draw. Read the message carefully and leave the last two options unchecked. Even if we don't want to consider how the datum will be transformed in this step, we still want to be warned every time we add a new layer with a different datum into the map display. Click on **Close** to dismiss the warning.

> If you accidentally check one of these options and want to display it again, go to `<ArcGIS_HOME>\Desktop10.4\Utilities` and double-click `AdvancedArcMapSettings.exe`. Select the **TOC/Data** tab and uncheck the following options: the **Skip datum check** and **Skip 'Unknown Spatial Reference'** warnings. Some of the options in the **ArcMap Advanced Settings** utility require administrator permissions. Make sure that your ArcMap application runs "As Administrator".

10. Repeat the last three steps to add the `ne_50m_populated_places_European1950.shp` and `ne_50m_populated_places_North_American1927.shp` shapefiles.

11. Add the last shapefile called `ne_50m_populated_places_Pulkovo42.shp` into the map display. A new warning message displays and tells you that spatial reference information is missing. Just because the dataset is missing a spatial reference does not mean it is not in a coordinate system. It just means that it has not been assigned or defined for the dataset so that ArcGIS can understand it and apply it to a proper **on-the-fly** transformation. Click on **OK**.

    At your current display scale (around 125,000,000), your shapefiles seem to perfectly overlap. We will take a closer look at your map and examine the difference between the point features represented by the fifth spheroids.

12. From the **Bookmark** menu, select **Detail 1- Paris (France)** to see a city at scale 1: 0.1, as shown in the following screenshot:

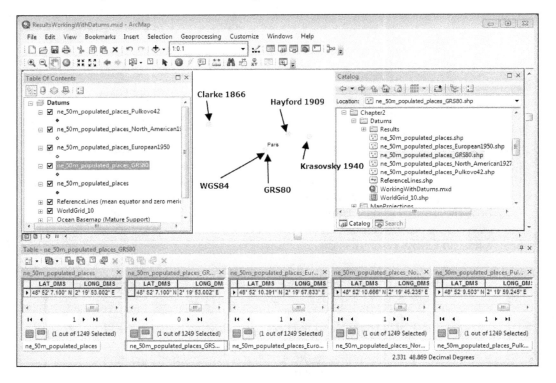

13. From the **Tools** toolbar, click the **Measure** tool and set its **Distance** to **Meters**. Measure the distance from the point represented on the Clarke1866 spheroid to the point represented on the WGS1984 spheroid. The difference between these two points should be around `197` meters. Depending on where you clicked, your value may be slightly different from the one shown in the previous screenshot.

14. Use the **Select Features** tool that is located on the **Tools** toolbar to select the five visible city points. Once selected, open **Attribute Table** for each layer by right-clicking the layer name in **Table Of Contents** and selecting **Open Attribute Table**. Click the tabs located at the bottom of the **Table** window to view the data associated with each layer. Depending on the geodetic datum, the geographic coordinates for the same city are different at the *seconds'* level.

To arrange the tables within the **Table** window so that each table is visible at the same time, click the tab of each table and drag it into the right-docked position. You can use the **Arrange Tables** commands on the **Table Options** icon, which is located at the upper-left corner of the **Table** window to change the place of an active table.

In the upcoming steps, we will choose the optimal on-the-fly datum transformation in order to fix or minimize the overlap problem. Follow these steps to start working with the ArcMap's on-the-fly datum transformations:

1. First, we will define the coordinate system for `ne_50m_populated_places_ Pulkovo42.shp`. In the **Catalog** window, right-click **ne_50m_populated_ places_Pulkovo42**`.shp` and navigate to **Properties | XY Coordinate System**. To select the current coordinate system, navigate to **Geographic Coordinate Systems | Europe** and select **Pulkovo1942**. Click on **Apply** and **OK**. In **Table Of Contents**, right-click the **ne_50m_populated_places_Pulkovo42** layer and navigate to **Properties | Source** to check whether the geographic coordinate system was updated. Click on **OK**.

   To correct the ArcMap's on-the-fly datum transformations, we have two options:

   1. Remove the layer and add it again.
   2. Use the **Data Frame Properties** dialog.

   In the next step, we will use both of these methods.

2. Remove the **ne_50m_populated_places_Pulkovo42.shp** layer from **Table Of Content** (right-click the layer and click **Remove**).

3.  From the **Catalog** window, drag it again into the map display. The **Geographic Coordinate Systems Warning** window will pop up. Click the **Transformations** button, and for the **Using (choices are sorted by suitability for layer's extent)** options, click the drop-down arrow and select the first transformation in the list: Pulkovo_1942_To_WGS_1984_16. Click on **OK** and **Close**. The ArcMap's datum transformation has reduced the point location shift. Unselect the features with the **Clear Selected Features** tool and zoom in close to the WGS84 point to take a closer look. Most probably you will need to zoom to a scale of 1:1.

 Before the use of the first option from the transformation drop-down list, we recommend that you confirm the suggested transformation by quickly studying the Esri document called geographic_transformations. pdf stored at <ArcGIS_HOME>\Desktop10.4\Documentation. This file lists all the supported datum transformations and areas for which they are suited.

Based on the evaluation of data extent and given datums, ArcGIS for Desktop will first list what it considers the most accurate transformation to use. However, the one ArcGIS recommends first may not be the best for your particular data. You may need to try multiple transformations before finding the one that works best with your data.

The information related to the selected transformation will be stored by ArcMap as part of **Data Frame Property**. This information will be used by your data frame when you will add another dataset having the same datum called D_Pulkovo_1942. Try to add ne_50m_populated_places_ Pulkovo1942.shp again, and you will notice that a new layer is added to **Table Of Contents** and displayed without any other geographic transformation warning.

In the succeeding steps, we will try to improve the datum transformation for the rest of the shapefiles.

4.  From the **Tools** toolbar, click the **Go Back To Previous Extent** tool to return to a best previous map scale. In **Table Of Contents**, right-click the **Datums** data frame and click **Properties**. Select the **Coordinate System** tab. Click the **Transformations** button to open the **Geographic Coordinate Systems Transformations** dialog, as shown in the following screenshot:

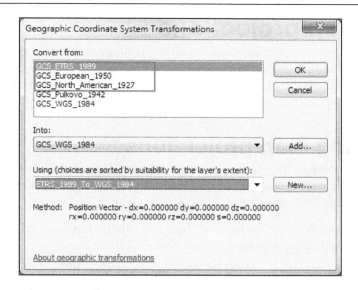

5. For **Convert from option**, select GCS_ETRS_1989. Leave **Into**: GCS_WGS_1984 unchanged. For the **Using** options, click the drop-down arrow and select the first transformation in the list: **ETRS_1989_To_WGS_1984**. *Do not click* on **OK**.

6. Repeat the last step to choose a datum transformation for GCS_Europe_1950 and GCS_North_American_1927. Click on **OK** and **Close**. Move the **Data Frame Properties** window in the ArcMap window so that you can see the map display area better. Click **Apply** and notice the results. Click **OK** to close the **Data Frame Properties** window.

7. Again, zoom in close to the point named WGS84 to take a closer look. Use the **Measure** tool to measure the small differences between these points.

   Please remember, ArcMap will temporarily calculate the coordinate values corresponding to the data frame's coordinate system through an on-the-fly datum transformation only for display purposes. This means that the coordinate reference system and stored coordinate values of the displayed shapefiles remain unchanged. If you want to permanently change the coordinate reference system of the shapefile, you would need to use the **Project** tool from **ArcToolbox | Data Management | Projections and Transformations** to project the data.

8. Save the map document at <drive>:\LearningArcGIS\Chapter2\Datums as MyWorkingWithDatums.mxd. Click on **File** and select **Exit** to close ArcMap.

   You can find the results of this exercise at <drive>:\LearningArcGIS\Chapter2\Datums\Results.

# Using map projections

This section does not undertake covering all the fundamental concepts of map projections because this subject is very well-documented in many recent books, and it is beyond the scope of this book. Instead, we will help you explore the most used map projections using the ArcMap application.

# Classifying map projections

Map projection refers to the mathematical transformation of the earth's three-dimensional surface (ellipsoid or sphere) into a two-dimensional surface (a flat one) through an intermediate-developable surface, such as cylinder, cone, or plane.

Map projections can be classified according to the distortions caused by converting the earth's surface to the projection, as follows:

- **Conformal projection**: This distorts the areas and distances but preserves the local directions and feature's shapes. This is commonly used for topographic and cadastral maps, mainly at a large scale, and navigation or route planning maps.

- **Equal Area projection**: This preserves the area of smaller and local features but instead distorts shapes or any combinations thereof. This is used mainly for thematic maps (for example, land use or statistical maps).

- **Equidistant projection**: This preserves the distances along certain meridians or parallels. This is suitable for flight distance maps.

Based on the graticule orientation (meridians and parallels), the map projections can be classified as direct (normal or polar), transverse (equatorial), and oblique. A direct *aspect* of map graticule is when the pole of the coordinate system (generically called Q) coincides with the geographical pole North (P), which means that the latitude of pole Q is $\varphi = 90^0$ (*Bugayevskiy, Map Projections: A Reference Manual, 2002*). The transverse aspect of map projection is when the latitude of the pole Q is $\varphi = 0^0$. When the latitude of the pole Q is $0^0 < \varphi < 90^0$, we have an oblique aspect of map projection.

Depending on the developable surfaces, the map projections can also be classified as the following:

- Cylindrical projections
- Conic projections
- Azimuthal projections

# Comparing map projections

Before we start exploring the map projections, we would like to recommend the following free sources to you:

- D2.8.I.1 Data Specification on Coordinate Reference System-Technical Guidelines, Infrastructure for Spatial Information in Europe, 2014, which can be found at `http://inspire.ec.europa.eu/documents/Data_Specifications/INSPIRE_DataSpecification_RS_v3.2.pdf`

- Map Projections for Europe, Institute for Environment and Sustainability, Join Research Center, 2001, which can be found at `http://ec.europa.eu/eurostat/documents/4311134/4366152/Map-projections-EUROPE.pdf`

- A Guide to coordinate systems in Great Britain, Ordnance Survey, 2015, which can be found at `https://www.ordnancesurvey.co.uk/docs/support/guide-coordinate-systems-great-britain.pdf`

In this section, we will explore the parameters, aspect, and deformations of twelve map projections. We will use the `Admin 0-Countries` dataset, which was downloaded from `www.naturalearthdata.com`. Follow these steps to start comparing the map projections using the ArcMap application:

1. Start ArcMap, and open the existing map document `Comparing Map Projections.mxd` from `<drive>:\LearningArcGIS\Chapter2\MapProjections`.

2. In **Table Of Contents**, the `ne_50m_admin_0_countries` layer contains all countries in the world at scale 1:50,000,000. The `TissotIndicatrices` layer stores some geodesic circles that will substitute for Tissot's indicatries and will help us visualize the distortions across the map better.

   The Tissot's indicatries are infinitely small geodesic circles on the ellipsoid that are projected on to the map projection as infinitely small ellipses. The shape and size of the ellipse of distortion (**indicatrix**) will show us how much the scale has changed and in which direction.

3. In **Table Of Contents**, right-click the **GCS_WGS_1984** data frame, select **Properties**, and click the **Coordinate System** tab. You can see the parameters of the current coordinate reference system, which is `GCS_WGS_1984`.

4. In the coordinate system list, expand the `Layers` folder. The `Layers` folder contains a single coordinate system that tells you that the layers in the data frame have the same Geographic Coordinate (Reference) System as the data frame. Don't close the **Data Frame Properties** dialog window.

To explore the map projection, we will manually change the data frame's coordinate reference system. ArcMap will automatically project all on-the-fly layers to match the new projected coordinate system that we have set. In this exercise, projecting the data frame is a good choice for us because we are only interested in a visual analysis of map projection's aspect and distortions. Next, we will work with cylindrical projections.

# Cylindrical projections

Cylindrical projections are used for world maps, but they are most suitable for regions close to the Equator and for countries, which are extending along the meridians (north-south).

The first example of cylindrical projection is the **Plate Carrée** projection. It is an equidistant map projection that preserves the distances, along the meridians, undistorted. The map projection has a normal aspect, which means that meridians and parallels are straight and equally spaced. Meridians and parallels intersect at right angles:

1. Let's start by applying the Plate Carrée projection. In the **Data Frame Properties** window, navigate to **Projected Coordinate Systems | World** and select WGS 1984 Plate Carree.

2. In the **Current Coordinate System** section, you can see the parameters of projection, the used datum, and the name and parameters of the spheroid. Click on **OK**.

3. Inspect the shape of the ellipses of distortion across the map. Along the Equator, the shape and the area of circles do not vary and are true (which means that the scale is true).

The second example of cylindrical projections is the **World Mercator** projection. It is a conformal map projection with a normal aspect that preserves the angle and distorts the area, distance according to the latitude. The Polar Regions have the biggest distortion of area. For example, Greenland (North) looks bigger than the African continent, or Antarctica (South) looks bigger than the rest of the continents, as you may see in the following screenshot:

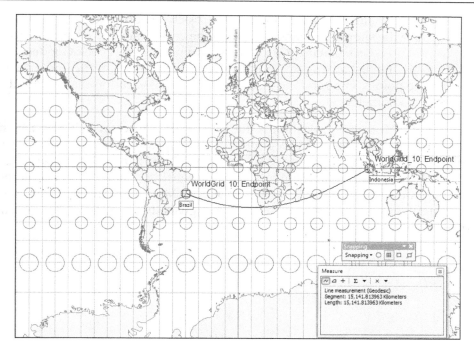

Follow these steps to select the **World Mercator** projection:

1. In **Data Frame Properties | Coordinate System**, navigate to **Projected Coordinate Systems | World** and select WGS 1984 World Mercator. Click on **OK**.

2. Inspect the shape of the ellipses. You may notice that all geodesic circles are projected as circles that vary in size, which means that feature's shapes are not distorted but the area is increasing towards the poles. Along the Equator, the scale of the map is true.

3. Let's measure the distance from Brazil to Indonesia in the Mercator projected coordinate system. First, from the **Customize** menu, click **Toolbars** and check **Snapping**. On the **Snapping** toolbar, select only **End Snapping**. From the **Tools** toolbar select the **Measure** tool. Click the **Choose Units** black down arrow, and select **Distance | Kilometers**.

4. Click the **Choose Measurement Type** arrow, and select **Planar**. With the **Measure Line** button selected, first click on the red circle next to Brazil and secondly double-click Indonesia, as shown the previous figure. The distance from Brazil to Indonesia is around 15,742 kilometers.

5.  Even if your current data frame is in a projected coordinate system, you can measure the shortest distance based on the curved surface of the spheroid WGS84 (geodesic distance). From **Measurement Type**, check **Geodesic**. Measure the distance again and please notice the shape of the line between the two points. The geodetic distance from Brazil to Indonesia is around `15,141` kilometers.

A version of the Mercator projection is the **Transverse Mercator ETRS89-TMzn,** which is recommended in the EU region by the European Commission (EC) for conformal pan-European mapping at scales larger than 1:500,000 (for example, 1:100,000, 1:50,000, 1:10,000). In the following screenshot, `ETRS89-TM31` centered on 30 E longitude of origin has a transverse aspect. Meridians and parallels are no longer straight, except the central meridian that gives the direction of true north, as shown in the following screenshot:

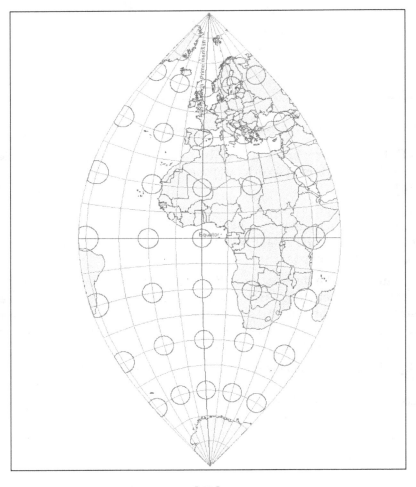

Follow these steps to select the **Transverse Mercator ETRS89-TM31** projection:

1. Navigate to **Data Frame Properties | Coordinate System**, expand **Projected Coordinate Systems | UTM | Europe**, and select **ETRS_1989_ETRS-TM31**.

2. Check the parameters of projection and the datum. The datum is different than D_WGS_1984, so you should expect a warning message. Click on **OK**. If you get the transformation warning message, click the **Transformation** button and select the first suggested transformation methods, as we learned in the previous exercise. Inspect the results. Notice that the scale is true along the central meridian.

The last example of cylindrical projections is the **Lambert Cylindrical Equal Area (World)** projection. It preserves the areas but distorts the shape. The shape and size of the countries are true close to the Equator and significantly distorted near the Polar Regions:

1. In the **Data Frame Properties | Coordinate System**, type cylindrical and click the **Search** button.

2. Select **Cylindrical Equal Area (World)** from **Projected Coordinate Systems | World**. Click on **OK**. Inspect the result.

Next, we will work with conic projections.

# Conic projections

Conic projections are most suitable for the mid-latitude regions (around $6^0$ extent north-south) and for countries that extend along the parallels (east-west). Map distortions are independent of longitudes.

The first example of conic projection is the **Europe Equidistant Conic** projection with standard parallels at $43^0$ and $62^0$ N. It is an equidistant map projection that preserves the distances along all meridians and along the standard parallels. The map projection has a normal aspect with straight meridians, which converges towards the pole (projected as a circular arc) and with equally-spaced parallels plotted as concentric circular arcs:

1. Navigate to **Data Frame Properties | Coordinate System**.

2. Expand **Projected Coordinate Systems | Continental | Europe**, and select **Europe Equidistant Conic**. Click on **OK**. Inspect the result.

A second example of conic projection is the **Lambert Conformal Conic** projection (**ETRS89-LCC**) with standard parallels at $35^0$ and $65^0$ North. It is recommended in the EU region by the European Commission (EC) for conformal pan-European mapping at scales smaller or equal to 1:500,000 (for example, 1:750,000, 1:5,000,000). The map projection has a normal aspect with straight, equally-spaced, meridians, which converge towards the pole (the point of the cone), as shown in the following screenshot:

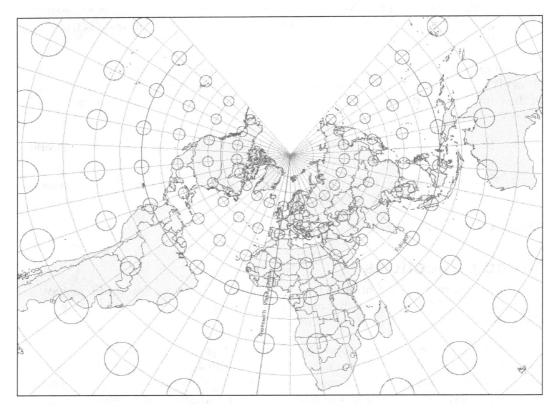

Follow these steps to select the **ETRS 1989 Lambert Conformal Conic** projection:

1.  Navigate to **Data Frame Properties | Coordinate System**.

2.  Expand **Projected Coordinate Systems | Continental | Europe**, and select **ETRS 1989 LCC**. Click on **OK**.

3.  Inspect the result. Along the two standard parallels the shape and size (scale) are not distorted.

    In the next step, we will learn how to change the aspect of the projection.

4. Double-click the data frame, and select the **Coordinate System** tab. In the coordinate system list, double-click **Europe Lambert Conformal Conic ETRS89-LCC**, to open the **Projected Coordinate System Properties** dialog. Change the name of the projection to ETRS_1989_LCC_sp=0, and for **Standard Parallel 1** and **2**, type 0, as shown in the following screenshots:

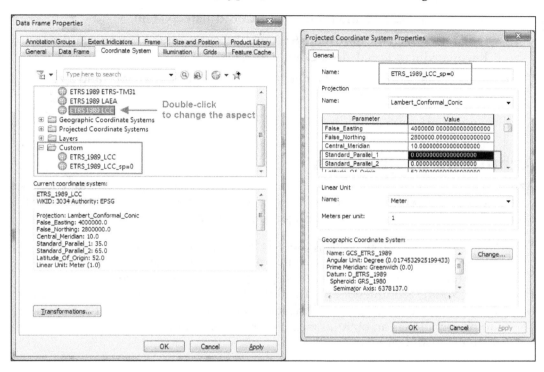

5. ArcGIS uses the new parameters to redraw the map projection using the Lambert Conformal Conic formulae. By setting standard parallels at $0^0$ (Equator), you will obtain a projection similar to Mercator cylindrical projection.

6. Let's change the standard parallels of the Europe Lambert Conformal Conic ETRS89-LCC to $90^0$ (North Pole) again. You will obtain a projection similar to a polar stereographic projection (Source: J. Iliffe & R. Lott, 2008).

The last example of conic projection is the **Europe Albers Equal Area Conic** projection with standard parallels at $43^0$ and $62^0$ North. It preserves the areas but distorts the shape near the Polar Regions. The map projection has a normal aspect with straight, equally-spaced, meridians, and converges towards the pole (projected as a circular arc), and intersects the parallels at right angles:

1. Navigate to **Data Frame Properties | Coordinate System**.

2. Expand **Projected Coordinate Systems | Continental | Europe**, and select **Europe Albers Equal Area Conic**. Click **OK**.

3. Inspect the result. Along the two standard parallels, the shape and size (scale) are not distorted.

Next, we will work with azimuthal projections.

# Azimuthal projections

Azimuthal (planar) projections are most suitable for the polar regions and for countries with an approximately circular form. These projections offer minimum distortions of distance and only the center point of projection is free from any distortion.

The first example of azimuthal projections is the **Lambert Azimuthal Equal Area** projection (**ETRS89-LAEA**) with the center of projection (projection pole) at $52^0$ N latitude and $10^0$ E longitude. It is a non-perspective azimuthal, equal-area, oblique projection recommended by the European Commission (EC) for true area representation of the EU region at all scales, spatial analysis, and report activities. The map projection has an oblique aspect with non-straight meridians that converges towards the pole and circular parallels.

Only the central meridian (10⁰ E longitude) is a straight line, as you can see in the following screenshot:

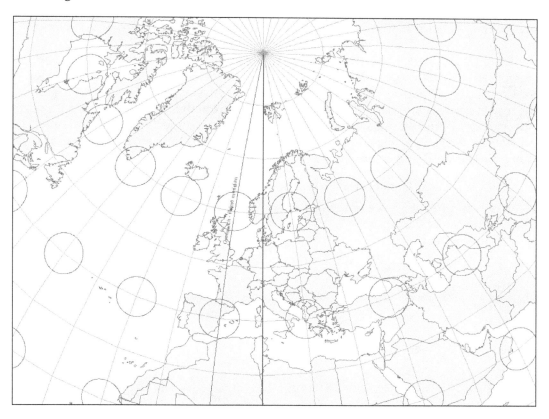

Follow these steps to select the **Lambert Azimuthal Equal Area** projection:

1. Navigate to **Data Frame Properties | Coordinate System**.
2. Expand **Projected Coordinate Systems | Continental | Europe,** and select **ETRS 1989 LAEA**. Click **OK**.
3. Inspect the result.

The second example of azimuthal projection is the **Stereographic Romania Stereo70** projection with the center point of projection at $46^0$ N latitude and $25^0$ E longitude (central meridian). It is a perspective azimuthal, conformal map projection that preserves the direction from a point to every other point. The map projection has an oblique aspect with non-straight meridians and parallels, except the central meridian ($25^0$ E), which is a straight line. The distances and areas are distorted. The stereographic (conformal) map projections are often used by small circular countries for large scale topographic and cadastral maps. The surveying engineers can compute the measurements (for example, horizontal angles) from the field directly into the map projection by reducing the distance and directions to the projection plane without calculating the geographic coordinates. Follow these steps to select the **Stereographic Romania Stereo70** projection:

1. Navigate to **Data Frame Properties | Coordinate System**.

2. Expand **Projected Coordinate Systems | National Grids | Europe**, and select **Pulkovo 1942 Adj 1958 Stereo 1970**. Click **OK** to obtain the following results:

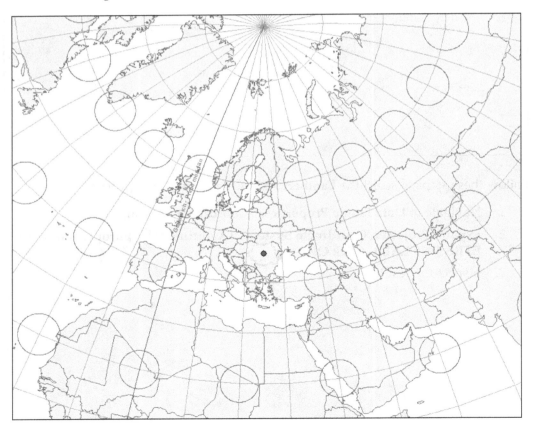

The last example of azimuthal projection is the **North Pole Lambert Azimuthal Equal Area** projection with the latitude of pole (center of projection) Q = 90⁰ N. It is a non-perspective azimuthal projection, which preserves the distances on the radial direction from the pole undistorted. The map projection has a normal aspect with straight meridians radiating from the pole (projected as a circular arc):

1. Navigate to **Data Frame Properties | Coordinate System**.

2. Expand **Projected Coordinate Systems | Polar,** and select **North Pole Lambert Azimuthal Equal Area**. Click **OK**. Notice how the distance between the parallels decreases towards the Equator.

3. In the last step, in the coordinate system list, try to find two modified projections, which are commonly used map projections for world maps at small scales: **Bonne** (pseudo-conic equal-area), and **Robinson** (pseudo-cylindrical).

4. Save the map document as MyMapProjections.mxd at <drive>:\ LearningArcGIS\Chapter2\MapProjections. Close ArcMap.

You can find all these maps saved as different data frames in the map document called MapProjections.mxd at <drive>:\LearningArcGIS\Chapter2\ MapProjections.

 To activate a data frame, right-click the data frame and click **Activate** or press the *Alt* key and select the data frame in **Table Of Contents**.

# Transforming coordinate reference systems

In this section, we will learn how to use correctly the coordinate reference system (CRS) transformation tools in ArcGIS. There are two ways of changing a CRS, as follows:

- Coordinate transformation (for CRSs with different datum)

- Coordinate conversion (for CRSs with common datum)

To perform both or one of the coordinate operations mentioned earlier, you should use the ArcToolbox tool called **Project**. The **Project** tool performs conversion from one CRS to another, based on the same common datum (for example, geographic coordinate system converts to a projected coordinate system). It also performs transformations between two CRS which have different datum, even if CRS is geographic or projected.

In the next exercise, we will change the map projection of a shapefile from the **Lambert Conformal Conic 1972** based on the `D_Belge_1972` datum to the **Lambert Conformal Conic 2008** based on the `D_ETRS_1989` datum, as shown in the following screenshot:

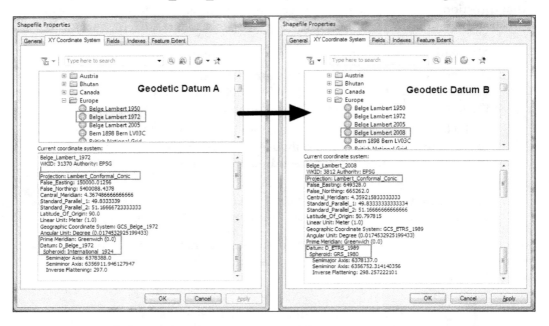

To change the shapefile's map projection, we will use the **Project** tool from ArcToolbox. The **Project** tool changes the coordinate reference system for a given dataset through a conversion or/and transformation, as shown in the following figure:

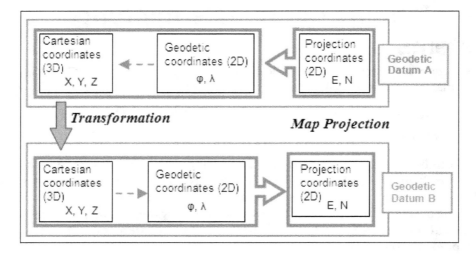

The figure above displays how the **Project** tool changes the coordinates between reference systems based on different datums:

- In Map projection, 2D projection coordinates are converted to 2D geodetic coordinates on the spheroid using map projection formulae
- Conversion from 2D geodetic coordinates to 3D Cartesian coordinate
- Datum transformation using the seven-parameter transformation method from Datum A (BD72 - Reseau National Belge 1972) to Datum B (ETRS89 - European Terrestrial Reference System 1989)
- Conversion from 3D Coordinate system to 2D geodetic coordinates
- In map projection, 2D geodetic coordinates are projected on the plane surface by the map projection formulae

Follow these steps to perform a data projection using ArcToolbox from ArcMap application:

1. Start ArcMap, and open the existing map document, `TransformCRS.mxd`, from `<drive>:\LearningArcGIS\Chapter2\Transformation`.
2. In **Table Of Contents**, the `Countries_Lambert1972` layer contains all countries in the world at scale 1:50,000.
3. From the **Standard** toolbar, click **ArcToolbox** and **Catalog**. In the **Catalog** window, navigate to `..\Chapter2\Transformation` and right-click the **Countries_Lambert1972** shapefile. Inspect the details related to **XY Coordinate System**. Click on **OK**.

4.  In the **ArcToolbox** window, navigate to **Data Management Tool |
    Projections and Transformations** and double-click the **Project** tool
    to open the dialog box, as shown in the following screenshot:

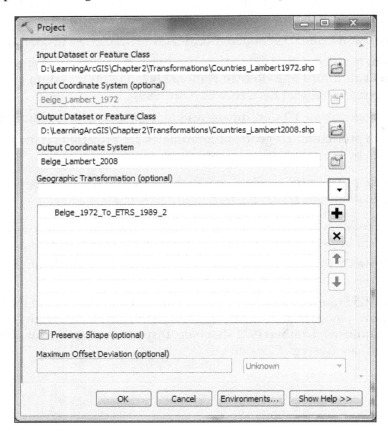

5.  For **Input Dataset or Feature Class**, select the `Countries_Lambert1972` layer
    from the drop-down list.

6.  For **Output Dataset or Feature Class**, click the yellow icon and navigate to
    `<drive>:\LearningArcGIS\Chapter2\Transformation`. For **Name**, type
    `Countries_Lambert2008`. In the **Save as** type, keep a single option available:
    **Feature classes**. ArcMap will know to save the results of projection into a
    new shapefile format and the reason is that you have chosen a shapefile
    format as input and the location of output is a folder. Click **Save**.

7.  For **Output Coordinate System**, navigate to **Projected Coordinate System
    | National Grids | Europe** and select `Belge Lambert 2008`. Inspect the
    details. Click on **OK**.

8. Now, it is time to choose the suitable transformation between two datums: `D_Berge_1972` and `D_ETRS_1989`. For **Geographic Transformation**, select the first option suggested by ArcMap called **Belge_1972_To_ETRS_1989_2**.

9. Before we start running the projection, let's check the transformation by studying the Esri document called `Geographic_tranformation.pdf`. Hint: take a look at `<ArcGIS_HOME>\Desktop10.4\Documentation`.

10. Open the PDF file. In **Table 1: Geographic (datum) transformation: well-known IDs, accuracies and area of use**, on page 7, you will find two transformations: `Belge_1972_To_ETRS_1989_1` with the `1652` code and a 1 meter precision and `Belge_1972_To_ETRS_1989_2` with the `15928` code with a 0.2 meter precision. The `15928` transformation seems to be the most accurate for our data.

11. In **Table 5: Geographic (datum) transformation: Coordinate frame (CF) and position vector (PV) methods**, on page 61, we can read some details about the `Belge_1972_To_ETRS_1989_2` datum transformation with the `15928` code, method use (CF), and the transformation parameter values of the seven-parameter similarity transformation method (three translations, three rotations, and one scale factor).

12. Let's double check the transformation method using the following URL: `epsg-registry.org`. Select the **retrieve by code** tab. In the upper-left corner, type the `15928` value in the **Code** section and click **Retrieve**.

13. Click the **view** link next to the **BD72 to ETRS89 (2)** transformation to see the details. Check the transformation parameter values. To find out more about the source coordinate reference system, expand the **Source CRS [Belge 1972]** section. To find more about the datum used, expand the **Geodetic Datum [Reseau National Belge 1972]** subsection. Next, expand the **Ellipsoid [International 1924]**. The **Aliases** section said the International 1924 ellipsoid is based on the general parameters of the Hayford 1910 ellipsoid.

14. Continue to explore the **Target CRS** section to find out more about the **ETRS89** coordinate reference system.

15. Let's return to the **Project** window and click on **OK**.

16. Add the `Countries_Lambert2008` shapefile into the map display. The **Geographic Coordinate Systems Warning** will be displayed.

17. Please remember the last exercise when we first ignore the warning message of the data frame. To see the shift between the two different datums, click **Close**. At your current display scale (around 100,000,000), your layers are perfectly overlapping.

18. Let's take a closer look at our map. From **Bookmarks**, select **Detail 1**. Repeat the step for all bookmarks. At scale 1:100,000, the differences among the layers should be visible. Use the **Measure** tool to measure the difference, as shown in the following screenshot:

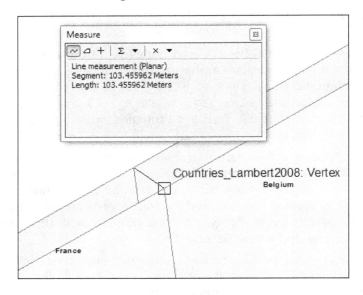

19. You may notice that Data Frame's Coordinate Reference System is `Belge Lambert 1972`. To check this, in **Table Of Contents**, double-click the `Layers` data frame to open the **Data Frame Properties** window and click the **Coordinate System** tab.

20. You can change **Coordinate Reference System** of the current data frame to Belge Lambert 2008 and apply the on-the-fly transformations, or you can open a new map document and add first the new projected `Countries_Lambert2008` shapefile into the map display; after this, you may add all the other layers as you wish. We recommend you to use the second option if you want to continue to work with the Belge Lambert 2008 map projection.

 If you want to check the coordinate values for the Belge Lambert 2008 projection in your projected dataset, please refer to http://www.ngi.be/NL/NL2-1-9.shtm.

21. Save the map document as `MyTransformCRS.mxd` at `<drive>:\ LearningArcGIS\Chapter2\Transformation`. Close ArcMap.

You can find the results of this exercise at `<drive>:\LearningArcGIS\Chapter2\ Transformation\Results`.

# Summary

In this chapter, you identified the main components of a coordinate reference system (CRS) in ArcMap. You also learned how to integrate datasets with different CRS in the same map using the "on-the-fly" datum transformation only for the visual analysis.

You used ArcGIS Online to add a predefined basemap to your map document.

You explored the major categories of map projections using a single dataset through the same ArcMap "on-the-fly" process.

You also learned how to assign a CRS to a dataset and how to correctly use the CRS transformation tools in ArcGIS using the Esri map documentation.

In the next chapter, you will learn how to create, document, and populate a geodatabase format with data.

# 3

# Creating a Geodatabase and Interpreting Metadata

In this chapter, we will explore different vector data formats. We will organize the spatial datasets acquired from external resources in a geodatabase. We will learn the basics of how to design a file geodatabase and create metadata.

By the end of this chapter, you will learn the following topics:

- Creating a file geodatabase
- Importing existing data to a file geodatabase
- Documenting a geodatabase using metadata

## Creating a geodatabase

The **geodatabase** is an Esri data storage format that can be used in the ArcGIS platform. There are three types of geodatabases: **Personal Geodatabase**, **File Geodatabase**, and **Multiuser Geodatabase** or **Enterprise Geodatabase**.

A personal geodatabase is suited for a single user or small workgroups, which create and manage small-sized databases. It can be read by multiple users at the same time but edited by one person at a time. A personal geodatabase is stored in Microsoft Access data files, and it has the .mdb file extension. The maximum size of a personal geodatabase is 2 GB (gigabyte), but the performance slows the larger the database becomes, especially when the size goes over 500 MB. Currently, the personal geodatabase is not supported by ArcGIS Pro, ArcGIS for Server, Portal for ArcGIS, and ArcGIS Online.

A file geodatabase is recommended for the ArcGIS for Desktop users over a personal geodatabase. A file geodatabase is suited for single users or small to medium workgroups, which create and manage small to medium-sized databases. It supports multiple editors at the same time, as long as they are editing different feature classes or tables. A file geodatabase is a collection of spatial or nonspatial datasets stored in a filesystem folder that works across the operating system. There is no limit to the size of the file geodatabase, but the maximum size of an individual data file stored in a file geodatabase is up to 1 TB (terabyte). Throughout this book, we will work with the file geodatabase format.

A multiuser geodatabase requires the use of the ArcSDE technology, and it is used by medium-sized departments or larger organizations that work and manage large-sized databases. Multiple users can view and edit a multiuser geodatabase at the same time. There are three types of multiuser geodatabases: **Desktop (SQL Server Express)**, **Workgroup (SQL Server Express** with the .mdf file extension), and **Enterprise (Oracle, SQL Server, DB2, Informix**, and **PostgreSQL)**. The number of editors and the size of database depend on the **database management system (DBMS)** used. The workgroup and enterprise geodatabases require the ArcGIS for Server, which is a separate product by Esri.

 If you want to know more about the advantages of using a geodatabase, please refer to *Learning ArcGIS Geodatabase, Hussein Nasser, Packt Publishing, 2014*.

# Components of a geodatabase

A geodatabase is a storage container (as a database or file structure) that stores vector, raster, and tabular data along with custom toolboxes, address locators, and data validation tools, such as a topology or geometric networks.

In this chapter, we will work with the following three primary components of the geodatabases: feature classes (standalone at the geodatabase level or in a feature dataset), nonspatial tables (alpha and numeric data attributes), and feature datasets.

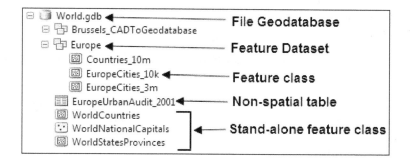

A feature represents a real-world object in a geodatabase, and it is the basic unit in the vector data model.

A feature class is a collection of features that share the same geometry type (point, line, or polygon), same characteristics (attributes), and a common spatial reference. The feature class table stores the geometry and attribute of features. The feature class table has a traditional format: *rows* and *columns*. Each row represents a single feature and the columns (fields) describe the attributes of the feature, as shown in the following screenshot:

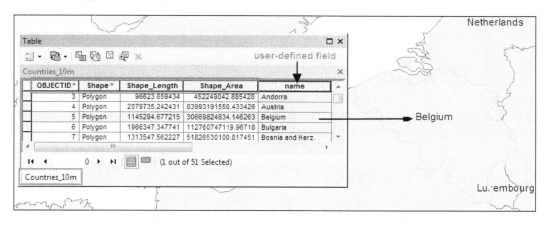

Depending on the geometry type stored, a feature class table contains the following four default attribute fields maintained by the geodatabase:

- **OBJECTID**: Its data type is `Object ID`, and it stores the unique identifier for each feature

- **Shape**: Its data type is `Geometry`, and it stores feature geometry (point, line, or polygon)

- **Shape_Length**: This attribute stores and maintains length values for line features or perimeter values for polygon features

- **Shape_Area**: This attribute stores and maintains area values for polygon features

The OBJECTID and SHAPE fields are automatically added to every new feature class in the geodatabase.

A feature class can be stored independently in a geodatabase as standalone feature classes, or they can be stored in a feature dataset. The *spatial reference* term refers to the following components: coordinate reference system (geographic or projected), spatial domain (min/max coordinate values of the common geographic extend), coordinate resolution, and coordinate tolerance.

A nonspatial table contains only attribute data without storing feature geometry. These tables are stored as standalone tables at the geodatabase level. They can be linked to other nonspatial tables or feature class tables.

A feature dataset groups feature classes of any geometry type, which are thematically related, and share the same spatial reference. A feature dataset allows you to maintain spatial relationships among feature classes through a geodatabase topology and build geometric networks and terrain datasets.

Your geodatabase is made of tables. Feature classes and nonspatial tables are tables in your **relational database management system (RDBMS)**. A feature dataset is not a container that stores feature classes in the geodatabase, though it may appear this way in ArcCatalog or the Catalog window. In a geodatabase, a feature dataset just relates different feature classes with a defined spatial reference, and this is stored as a row in a hidden internal table that is managed by ArcGIS for Desktop. Consequently, your feature class and table names must be unique in your geodatabase whether they are standalone or in a feature dataset.

# Creating a file geodatabase

A file geodatabase can be built, edited, and managed with ArcToolbox and with both the ArcCatalog and ArcMap applications using the context menu in the Catalog window. Follow these steps to start creating a file geodatabase using the ArcCatalog context menu:

1. Start the ArcCatalog application. Under **Folder Connections**, expand the <drive>:\LearningArcGIS\Chapter3 folder.

2. In **Catalog Tree** section, right-click on the **MyWorld** folder and under **New**, select **File Geodatabase** to create a new empty file geodatabase, as shown in the following screenshot:

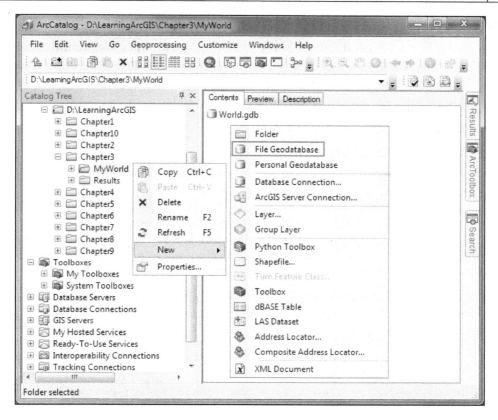

3. To change the default name, click on the new geodatabase, type World, and press the *Enter* key. Your World file geodatabase is ready to be populated with the existing data.

# Importing existing data

Importing existing data into a geodatabase is one of the two primary ways to build a physical geodatabase schema (geodatabase structure). The second way is to manually create the geodatabase structure and the data. In this section, we will combine both methods to build the structure of the geodatabase and populate it with existing data.

# Evaluating existing data

Before designing a file geodatabase and populating it with data, we should identify the source of the existing data and evaluate them. When we are evaluating existing data, we should answer the following questions:

- What is the format of the dataset?
- What is the geometry type of the dataset?
- What is the coordinate reference system?
- What is the reference scale (or resolution) of the dataset?
- What is the format of the initial dataset?

We will use four public datasets sources: Natural Earth (www.naturalearthdata. com), Eurostat (http://ec.europa.eu/eurostat), Urban Atlas from European Environment Agency (http://www.eea.europa.eu/data-and-maps/data/urban-atlas), and CAD Mapper (https://cadmapper.com).

From www.naturalearthdata.com/downloads, you can download all vector themes in the shapefile or geodatabase format, or you could download only the following two datasets at scale 1:10,000,000 (or 10 m):

- http://www.naturalearthdata.com/downloads/10m-cultural-vectors/10m-admin-0-countries/
- http://www.naturalearthdata.com/downloads/10m-cultural-vectors/10m-populated-places/

Unzip the archived shapefiles to <drive>:\LearningArcGIS\Chapter3\Data\Natural Earth, and explore the datasets using Windows Explorer and ArcCatalog applications, as shown in the following screenshot:

A shapefile needs at least three files to be displayed in ArcCatalog:

- `ne_10m_admin_0_countries.dbf`: This stores feature attributes in a dBASE format table

- `ne_10m_admin_0_countries.shp`: This stores feature geometry

- `ne_10m_admin_0_countries.shx`: This stores the index that links the records in the `dbf` and `shp` files

Even though you only see the `ne_10m-admin-0-countries.shp` file in ArcCatalog or the **Catalog** window in ArcMap, ArcGIS understands that all the other associated files work together to form a single shapefile.

 We have already downloaded and prepared the data from Natural Earth at `<drive>:\LearningArcGIS\Chapter3\MyWorld\Natural Earth`.

Let's inspect the shapefile properties, as follows:

1. Open ArcCatalog. In the **Catalog Tree** section, navigate to `<drive>:\LearningArcGIS\Chapter3\MyWorld\Natural Earth` and select the **ne_10m-admin-0-countries** shapefile.

2. In the **Preview** tab, examine the geometry and the records in the attribute table using the **Geography** and **Table** preview modes at the bottom of ArcCatalog application window.

3. Right-click on the **ne_10m-admin-0-countries** shapefile and click on **XY Coordinate System** in **Properties**. The current coordinate reference system of the dataset is the geographic coordinate system called `GCS_WGS_1984`. Click on **OK**.

From Eurostat, `http://ec.europa.eu/eurostat/web/gisco/geodata/reference-data/administrative-units-statistical-units`, download a dataset called URAU 2001.zip. Unzip the archive files to `<drive>:\LearningArcGIS\Chapter3\Data\Eurostat` and inspect the properties of the dataset using ArcCatalog application, as shown in the following screenshot:

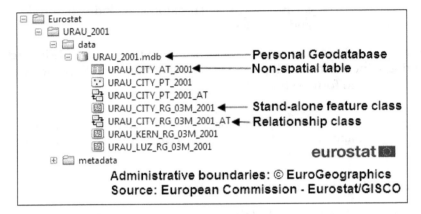

The URAU_2001 dataset contains spatial and statistical data for Europe. The reference scale of the dataset is 1:3,000,000 (or 3m). The dataset is stored in the personal geodatabase format, and the current coordinate reference system of the feature classes is the geographic coordinate system called GCS_ETRS_1989.

From European Environment Agency, `http://www.eea.europa.eu/data-and-maps/data/urban-atlas`, download two datasets called be0011_bruxelles.zip (from Belgium) and be0011_paris.zip (from France). Scroll to the bottom of the Urban Atlas webpage to read some additional information about the dataset's coordinate reference system. Unzip the archive files to `<drive>:\LearningArcGIS\Chapter3\Data\UrbanAtlas` and explore the properties of datasets using the ArcCatalog application. The two datasets are stored in the shapefile format and contain two European capitals: Brussels and Paris. The current coordinate reference system of the shapefile is `<Unknown>`. However, the **Additional Information** and **Metadata** sections from the top of the Urban Atlas webpage mention that the coordinate reference system is **LAEA/ETRS89**, which means that the Lambert Azimuthal Equal Area map projection with the D_ETRS_1989 datum. The reference scale of the dataset is 1:10,000.

From CAD Mapper, `https://cadmapper.com`, download the datasets called `brussels.zip` (from the Europe section). On the CAD Mapper webpage, you can read some details about the free city datasets. The datasets are stored in the DXF format, which is a **Computer-Aided Drafting (CAD)** format. Regarding the coordinate reference system, CAD Mapper mentioned the following: "The origin (0, 0) of the coordinate system is placed at the bottom-left of the area", as shown in the following screenshot:

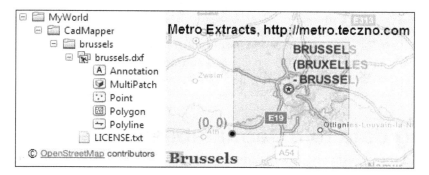

To summarize, we have three types of vector datasets (shapefile, personal geodatabase, and AutoCAD DXF) at different reference scales (small scale 1:10,000,000, 1:3,000,000, and medium scale 1:10,000) and with different coordinate reference systems (GCS_WGS_1984, GCS_ETRS_1989, and ETRS_1989_LAEA).

# Defining a geodatabase structure

When we are building a geodatabase structure, we should answer the following general questions:

- How many feature classes and feature datasets are in the file geodatabase?
- What is the coordinate reference system of standalone feature classes?
- What is the coordinate reference system of the feature datasets?
- What is the geometry type of feature class stored in the file geodatabase?

The following screenshot answers the previous questions:

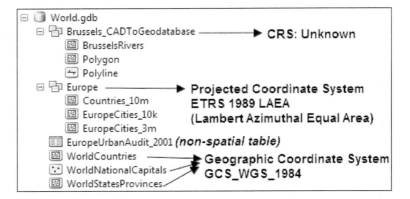

# Adding data to a geodatabase

We will add the previously downloaded data to the file geodatabase using different methods, such as **Import**, **Export**, and **Load data** tools from the ArcCatalog context menu and the **CAD to Geodatabase** tool from ArcToolbox.

 All the steps in this section need to be performed in a single session.

Follow these steps to start adding existing data to the World.gdb file geodatabase using the ArcCatalog application:

1. We will start by importing a shapefile and exclude some user-defined fields that are unnecessary for our geodatabase. In the **Catalog Tree** section, right-click on the **World.gdb** file geodatabase and select **Feature Class (single)** from **Import**. In the **Feature Class to Feature Class** window, set the tool parameters, as shown in the following screenshot:

2. For **Input Features**, select **ne_10m_admin_0_countries.shp** from `<drive>:\`
   `LearningArcGIS\Chapter3\MyWorld\Natural Earth`. For **Output Feature
   Class**, type `WorldCountries`. Under **Field Map**, remove all user-defined
   fields, except these four: **type**, **name**, **pop_est**, **gdp_md_est**, and **continent**.
   To remove the **scalerank (Short)** field, right-click on it and select **Delete**.
   Select the **featurecla (Text)** field and click on the **Delete** button. Click on
   **OK** to start running the tool.

3. In **Catalog Tree**, expand the **World.gdb** file geodatabase to see the new
   standalone feature class named **WorldCountries**. If you don't see the new
   feature class, right-click on the **World.gdb** file geodatabase and select
   **Refresh**. Right-click on the new feature class to open the **Feature Class
   Properties** window and click on the **Fields** tab. Note that the shapefile **FID**
   field was replaced by the **OBJECTID** field. The `WorldCountries` feature
   class has two new fields, **Shape_Length** and **Shape_Area**, which store the
   feature geometry.

4. Click on the **XY Coordinate System** tab. Note that spatial reference was
   imported along with the data, and in this case, the current coordinate
   reference system is **GCS_WGS_1984**. Click on **OK**.

Next, we will use the **Export** tool to add the national capitals from the
`ne_10m_populated_places.shp` shapefile to the `World.gdb` file geodatabase:

5. In the **Catalog Tree** section, right-click on the shapefile and select **To
   Geodatabase (single)** from **Export**. For **Output Location**, navigate to
   `<drive>:\LearningArcGIS\Chapter3\MyWorld`, select the `World.`
   `gdb` file geodatabase, and click on **Add**. For **Output Feature Class**, type
   `WorldNationalCapitals`. For **Expression (optional)**, click on the **SQL**
   button on the right to open the **Query Builder** dialog, as shown in the
   following screenshot:

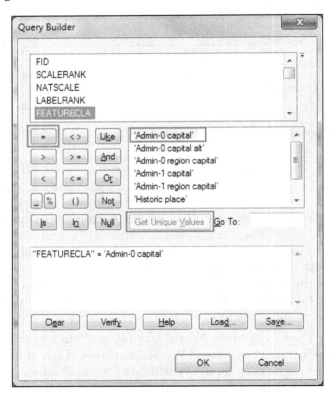

6.  To build the query expression `"FEATURECLA" = 'Admin-0 capital'`, double-click on `FEATURECLA` to add the field to the SQL text box. Click on the equal button (=) from the list of operators and click on the **Get Unique Values** button to get the list of city classes.

7.  In the list of unique values, double-click on **Admin-0 capital** to add the value to the SQL text box. Once you have built your expression, it should look like the one in the previous screenshot. Click on **OK** to close the **Query Builder** window.

8.  Under **Field Map**, keep only the `NAMEASCII` and `POP_MAX` fields.

> During import or export, you can rename these fields, or you can add your own new empty fields.

9.  Click on **OK**. After the tool finishes successfully, inspect the properties of the resulting standalone feature class.

In the following steps, we will import a feature class structure and its data from an XML workspace document file named `ne_10m_admin_1_states_provinces_geodb.xml`. This XML data exchange format is a geodatabase-friendly format that can be read by other applications:

10. In the **Catalog Tree** section, right-click on the **World** file geodatabase and select **XML Workspace Document** from **Import**. In the first panel, select **Data** to import both the schema and the data to geodatabase. For **Specify the XML source to import,** navigate to `..\MyWorld\Natural Earth`, select **ne_10m_admin_1_states_provinces_geodb.xml**, and click on **Open**. Click on **Next** and then click on **Finish**.

> To change the name of the feature class, right-click on the `ne_10m_admin_1_states_provinces_geodb` feature class, select **Rename**, and type `WorldStatesProvinces`.

11. Let's examine the properties of the resulting standalone feature class. Open the **Feature Class Properties** window and select **Fields** tab. We will keep only the `name` and `admin` fields. Please, remember that the required fields, such as `OBJECTID`, `Shape`, `Shape_Length`, and `Shape_Area`, are maintained by the geodatabase and cannot be deleted. To exclude a field from the attribute table, click on the small button to the left of the field name [step **(1)** in the next screenshot] and press the *Delete* key [step **(2)** in the next screenshot]. Repeat the steps for the rest of the user-defined fields, as shown in the following screenshot:

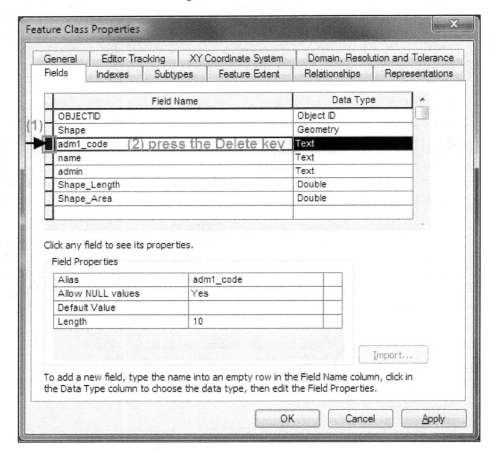

12. Click on **Apply** to apply these changes and on **OK** to close the **Feature Class Properties** window.

Until this point, you have created only the standalone feature classes.

Next, we will create a feature dataset to group the feature classes that store data only for the European continent. All feature classes in a feature dataset will share the same coordinate reference system, spatial domain, resolution, and tolerance:

13. In the **Catalog Tree** section, right-click on the **World.gdb** geodatabase, and select **New | Feature Dataset**. For **Name**, type Europe. Click on **Next**.

14. In the coordinate system list, expand the **Projected Coordinate Systems** folder. Navigate to **Continental | Europe** and select **ETRS 1989 LAEA**. Click on the **Add To Favorites** tool (yellow star button) to add the Lambert Azimuthal Equal Area projection in the **Favorites** folder. Make sure that ETRS 1989 LAEA remains selected in the **Favorites** folder. Click on **Next**.

15. In the coordinate system list, expand **Vertical Coordinate Systems | Europe**, and select **EVRF 2007**. Click on **Next**.

    The default value for **XY Tolerance** is one-thousandth of a meter (1 millimeter), which is the default linear unit of the projected coordinate system that is selected in the second panel.

     If you chose a geographic coordinate system (for example, GCS_ETRS_1989), you would have had the default **XY Tolerance** value equals to 0.000000008983153 degree, which is equivalent to 1 millimeter in the linear unit.

16. Accept the default **XY Tolerance**, **Z Tolerance**, and **M Tolerance** values. The **Accept default resolution and domain extent (recommended)** option is checked by default.

17. If you want to inspect the default values for **XY Resolution**, **Z Resolution**, and **M Resolution** and the domain extent for Z and M values, uncheck the default option, and click on **Next**. This panel allows you to change the default values if this is necessary. We will not modify these values. Click on **Finish**.

18. In the **Catalog Tree** section, right-click on the **Europe** feature dataset and select **Properties**. In the **Feature Dataset Properties** window, select the **Domain, Resolution and Tolerance** tab. Inspect the default values. You cannot modify the **Resolution** and **Tolerance** values for an existing feature dataset, but you can still modify the **XY Coordinate System** and **Z Coordinate System**. Changing **XY Coordinate System** and **Z Coordinate System** for a dataset doesn't imply a coordinate transformation for the data stored in it. Click on **OK**.

Next, we will combine data from the `be0011_bruxelles` and `be0011_paris` shapefiles (Urban Atlas) into a single feature class. Please, remember that the datasets of the Urban Atlas are already projected to Lambert Azimuthal Equal Area:

19. First, we will create an empty feature class in the `Europe` feature dataset. In the **Catalog** window, right-click on the **Europe** feature dataset and select **Feature Class** from **New**. In the **New Feature Class** window, for **Name**, type `EuropeCities_10k`. For **Type**, select **Polygon Features**. Leave the **Geometry Properties** options unchecked.

20. Click on **Next**. Accept **Default**, select the **Configuration Keyword** options, and click on **Next**.

21. The **OBJECTID** and **SHAPE** fields are automatically added to your new feature class. As we plan to store the Bruxelles and Paris cities in one feature class based on their common structure, we will import the field definitions from the `be0011_bruxelles.shp` shapefile. Click on the **Import...** button, navigate to `..Chapter3\MyWorld\Urban Atlas\be0011_bruxelles`, and select the `be0011_bruxelles` shapefile. Click on **Add**. Inspect the fields and click on **Finish**.

22. Your empty `EuropeCities_10k` feature class automatically inherits the spatial reference, domain, and XY resolution and tolerance of the feature dataset. Check this by opening **Feature Class Properties**.

23. We will load data from two shapefiles at once into `EuropeCities_10k` feature class. Right-click on the empty feature class, and select **Load Data** from **Load**. Click on **Next** in the first introductory panel.

24. For **Input data**, navigate to `..Chapter3\MyWorld\Urban Atlas\be0011_bruxelles` and select **be0011_bruxelles.shp**. Click on **Open**, and then click on **Add** in the **Simple Data Loader** panel. Navigate to `..Chapter3\MyWorld\Urban Atlas\be0011_paris` again and select `be0011_paris.shp`. Click on **Open** and then **Add** to add the second feature class to the list of source data, as shown in the following screenshot:

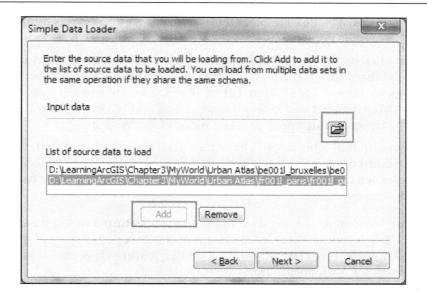

25. Click on **Next**. In the third panel, accept the default selected option.

26. Click on **Next**. This panel shows how the fields from the shapefiles will be mapped to the target fields (EuropeCities_10k). As a result of previous field definitions import, we expect to be a perfect match between source and destination fields. Click on **Next** twice and then click on **Finish**. Inspect the results in the **Preview** section.

In the next steps, we will learn about another way to add data into our World.gbd geodatabase. We will export a feature class named URAU_CITY_RG_03M_2001 and a nonspatial table named URAU_CITY_AT_2001 from the Eurostat personal geodatabase called URAU_2001:

27. In the **Catalog Tree** section, navigate to ..\MyWorld\Eurostat\URAU_2001\ data and expand the URAU_2001.mdb personal geodatabase.

28. Right-click on the URAU_CITY_RG_03M_2001 feature class and select **To Geodatabase (Single)** from **Export**. Set the following parameters:

    ° **Output Location**: Set this to ..\Chapter3\ MyWorld\World.gdb

    ° **Output Feature Class**: Set this to EuropeCities_3m_ETRS89

    ° **Expression (optional)**: Leave this empty

    ° **Field Map**: Keep the URAU_CITY_ID field and remove the Shape_ Length and Shape_Area fields

29. Click on **OK**. The `EuropeCities_3m_ETRS89` feature class is added to the `World.gdb` file geodatabase.

30. Right-click on the `URAU_CITY_AT_2001` nonspatial table and select **To Geodatabase (Single)...** from **Export**. For **Output Location**, select the `World.gdb` geodatabase. For **Output Table**, type `EuropeUrbanAudit_2001`. Under **Field Map**, inspect and keep all fields. Click on **OK**. Inspect the results using the **Geography** and **Table** preview modes from ArcCatalog.

    As a nonspatial table doesn't have a spatial reference associated with it, we cannot add the `EuropeUrbanAudit_2001` table to the `Europe` feature dataset. Instead, we can add the `EuropeCities_3m_ETRS89` standalone feature class to the feature dataset.

31. Before we start importing data to the `Europe` feature dataset, we should check the coordinate reference system of the `EuropeCities_3m_ETRS89` standalone feature class. Right-click on this feature class, and select **Properties | XY Coordinate System**.

    The feature class has associated a **Geographic Coordinate System** named `GCS_ETRS_1989` with the `D_ETRS_1989` datum. Even if the `Europe` feature dataset has a **Projected Coordinate System** named `ETRS_1989_LAEA` associated with it, both coordinate reference systems share the same datum called `D_ETRS_1989`. This means that the **Import** tool will need to project the 2D geodetic coordinates on the plane surface using the LAEA map projection formulae without any datum transformation.

32. Close the **Feature Class Properties** window. Right-click on the **Europe** feature dataset, and select **Import | Feature Class (Single)**. For **Input Features**, navigate to `..\Chapter3\ MyWorld`, select **EuropeCities_3m_ETRS89**, and click on **Add**.

33. For **Output Feature Class**, type `EuropeCities_3m`. Under **Field Map**, remove the `Shape_Length` and `Shape_Area` fields and keep the `URAU_CITY_ID` field. Click on **OK**.

34. To avoid storing duplicated data, right-click on the **EuropeCities_3m_ETRS89** standalone feature class and select **Delete**. Inspect the results.

Next, we will add only the European countries from the `WorldCountries` standalone feature class to the `Europe` feature dataset. Note that `WorldCountries` has a different CRS (with a different datum) than the `Europe` feature dataset. This means that the **Import** tool will apply a default datum transformation and a map projection:

35. Right-click on the **Europe** feature dataset, and select **Feature Class (Single)** from **Import**. For **Input Features**, navigate to `..\Chapter3\ MyWorld\World. gdb`, select **WorldCountries**, and click on **Add**.

36. For **Output Feature Class**, type `Countries_10m`. For **Expression**, build the following query expression, `"continent" = 'Europe'`.

37. Under **Field Map**, keep all the attribute fields. The `Shape_Length` and `Shape_Area` fields will be automatically superscript in the new feature class. Perimeter and area values will be calculated and updated for the Lambert Azimuthal Equal Area map projection. Click on **OK**. Inspect the results.

> If your feature class has a different datum than the feature dataset and you work with a large scale of data within a specific region, you should use the **Import** tool in conjunction with the **Project** tool from ArcToolbox.

Next, we will import a CAD file from CAD Mapper to the `World` geodatabase using the **CAD to Geodatabase (Conversion)** tool from ArcToolbox:

38. In the **Catalog Tree** section, navigate to `..\MyWorld\CadMapper\Brussels` and expand the **brussels.dxf** CAD file. Right-click on the CAD file and select **General** from **Properties**. The CAD coordinate reference system is undefined. In **CAD Feature Dataset Properties**, click on the **Coordinates** tab. The values from **CAD Dataset Extends** tell us that the feature coordinates are defined in a local coordinate system. Click on the **Details** tab to find out more about the structure of the DXF file. The `brussels.dxf` dataset contains only polygon and polyline features. Click on **OK**.

39. Inspect the geometry and attribute table of both the **Polygon** and **Polyline** geometries from the DXF format file. As you can see, the CAD feature classes have a lot of fields that store the visual characteristics of a CAD drawing, such as `Color`, `LyrColor`, `LineType`, `LineWT`, or `Thickness`. These fields are not relevant to our geodatabase, except the `Layer` field that stores the corresponding CAD layers.

40. From the **Standard** toolbar, select the **Search** tool. Select the **Tools** filter, type
    CAD, and click on the **Search** button. Several tools are returned. Double-click on
    **CAD to Geodatabase (Conversion)** to open the tool dialog. Set the parameters
    in the tool dialog, as shown in the following screenshot:

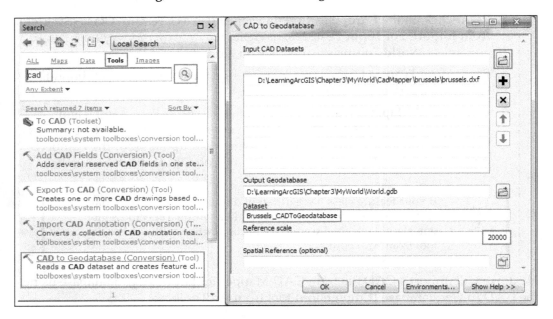

41. Click on **OK**. Note the progress bar from the bottom of the ArcCatalog
    application window.

42. The **CAD to Geodatabase** tool created a feature dataset called Brussels_
    CADToGeodatabase that stores the feature geometry in two different feature
    classes. Starting from these two feature classes and using the Layer field
    values, you may create new derivate feature classes. For example, you may
    create a new feature class that stores only the water from Brussels. Right-
    click on the **Polygon** feature class and select **To Geodatabase (Single)** from
    **Export**. Set the parameters in the tool, as shown in the following screenshot:

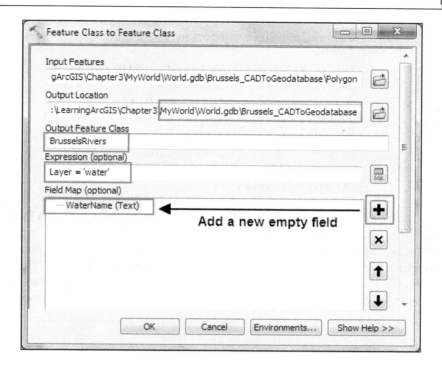

43. In the end, note that the `Brussels_CADToGeodatabase` feature dataset has no coordinate reference system defined. Use its **Feature Dataset Properties** dialog window to check this.

You can find the results of this exercise at `<drive>:\LearningArcGIS\Chapter3\Results\+Geodatabase`.

In *Chapter 5*, *Creating and Editing Data*, we will spatially adjust the feature classes using the **Spatial Adjustment** tool, and we will assign the `Brussels_CADToGeodatabase` feature dataset a coordinate reference system.

# Documenting a geodatabase using metadata

Metadata is commonly defined as *data about data*. Metadata provides information about the content, lineage (history and quality of the spatial dataset), conditions and limitations for the access and use of spatial data, and other characteristics of data. The metadata information helps users easily evaluate the quality of data and decide whether the dataset is appropriate for their projects.

For more information about different metadata standards, please refer to the following:

- ISO/TC211 standards (EN ISO19115, 19119 and 19139) at `http://www.isotc211.org/`.
- FGDC Geospatial metadata standards at `https://www.fgdc.gov/metadata/`.
- INSPIRE metadata standard at `http://inspire.ec.europa.eu/`.
- Dublin Core Metadata Initiative at `http://dublincore.org/`.

Regarding metadata, the INSPIRE Directive 2007/2/EC of the European Parliament and of the Council define metadata in Art. 3, paragraph (6) as the following: "information which describes spatial data sets and services and which allow their search, inventory and use." The following graphic shows us that the INSPIRE metadata profile is composed of ten elements. We have expanded only two of these elements for you to inspect the specific subelements of a metadata document:

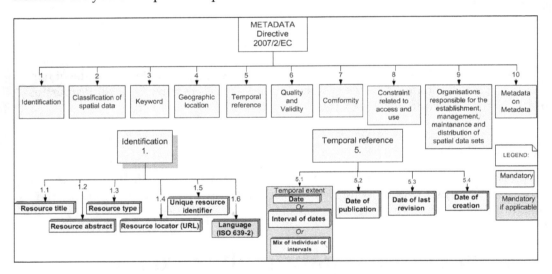

The most common file format used to distribute the metadata information is through a standardized XML format file. In ArcGIS for Desktop, you can view, create, edit, import, and export metadata using the metadata editor. ArcGIS for Desktop provides different metadata styles to view metadata, such as `North American Profile of ISO19115 2003`, `ISO 19139 Metadata Implementation Specification`, or `Inspire Metadata Directive`. ArcGIS for Desktop will automatically create and update a default metadata style named **Item Description**.

The **Item Description** metadata style provides basic metadata information, such as the resource title, resource type, and reference system of the data. Supplementary information, such as tags, purpose, description, resource constraints, or data quality should be manually updated by the user.

# Creating metadata

It's good practice to create metadata while you are adding or creating new data in the file geodatabase. Therefore, we will start using the metadata editing interface to create metadata in the ArcMap application:

1. Start ArcMap and click on **Cancel** in the **Getting Started** dialog to open a new map document.

2. First, let's set up the metadata style in ArcMap. From the **Customize** menu, select **ArcMap Options** and click on the **Metadata** tab. For **Metadata Style**, select the **ISO 19139 Metadata Implementation Specification** style from the drop-down list. Click on **OK**.

3. Open the **Catalog** window and click on the **Auto Hide** button to pin the window open, as shown in the following screenshot:

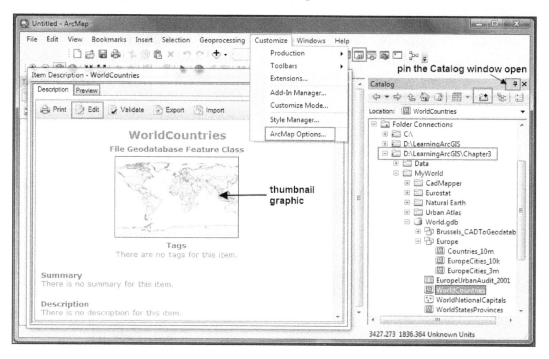

4. Use the **Connect To Folder** button to create a folder connection to `<drive>:\`
`LearningArcGIS\Chapter3`. Navigate to `..\Chapter3\MyWorld` and expand
the `World.gdb` geodatabase.

Now, we will create metadata for the `WorldCountries` standalone feature class
without adding it to **Table Of Contents**. Instead of adding the feature class to the
map, we will use **Item Description** to view the geometry and attributes and review
the metadata:

5. Right-click on **WorldCountries** and select **Item Description**.

6. Scroll down to inspect all the metadata information that was generated
automatically by ArcGIS for the `WorldCountries` feature class. Note that
some characteristics of your dataset are missing, such as tags, description,
use limitations, or scale range. These characteristics of the dataset should
be updated by the user.

7. Let's add a thumbnail graphic. Select the **Preview** tab and inspect the features
with the **Identify** tool. You can examine the feature class attribute table by
selecting **Table** from the **Preview** drop-down list.

8. In the **Geography** preview mode, click on **Create Thumbnail** tool, which is
next to the **Identify** tool, to create a snapshot graphic. To see the thumbnail,
click on the **Description** tab again.

9. To start editing the metadata, click on **Edit** to open the metadata editor.
Along the left side of the window is the list of metadata elements organized
in three main sections: **Overview**, **Metadata**, and **Resource**. As you can see
in the following screenshot, there are several sections that are marked with
a red **X**:

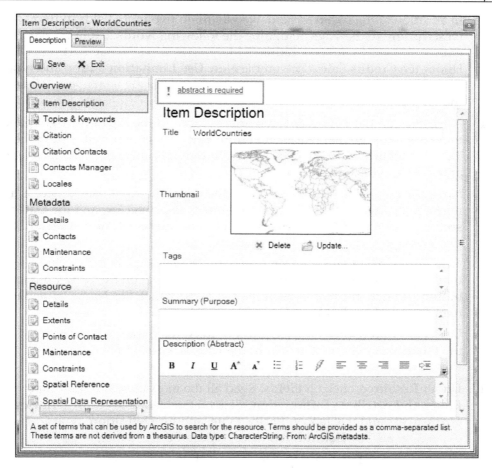

10. These sections contain incomplete elements and subelements of metadata, which are mandatory in the standard *ISO 19139 Geographic information-Metadata-XML schema implementation* (for example, **Item Description**, **Topics & Keywords**, and **Citation**).

11. Let's complete the **Item Description** section in **Overview**. At the top of the right panel, there is a message that tells us that an **abstract is required**. Scroll down to view the **Description (Abstract)** field marked red. Fill in the red field with the following text: The WorldCountries is a dataset that contains 247 countries in the world. Dataset was originally built by the Natural Earth as a public dataset at scale 1:10 million. Press the *Tab* key. Note that field turns to white, the message at the top has disappeared, and the **Item Description** label on the left is marked with a green tick.

12. In addition of the required field, we will fill in two more fields in **Item Description**. In the **Tags** field, type the following words: country, natural earth, and world. In the **Credits** field, type Made with Natural Earth. Under the **Credits** field, click on the **New Use Limitation** button and type: Free vector and raster map data @ naturalearthdata.com. Press the *Tab* key.

13. Next, click on the **Topics & Keywords** section. At the top of the right panel, there is a message that tells us that **topic category is required**. Under the **Topic Categories** section, click on the **Boundaries** checkbox. The message at the top has disappeared.

14. Next, click on the **Citation** section. At the top of the right panel, there is a message that tells us that **at least one date is required**. Expand the **Dates** section, click on the **Created** calendar button, and select **August 24, 2015**. The message at the top has disappeared.

15. Next, under **Metadata**, click on **Contacts**. The message at the top indicates that **at least one metadata contact is required**. Click on the **New Contact** button and note the messages at the top of the panel. For **Name**, type your name. For **Role**, select **User** from the drop-down list.

16. Under **Resource**, click on **Distribution**. Read the message and expand the **Distribution Format** arrow. In the **Format Version** field, type 10.4. Press the *Tab* key.

17. Under **Resource**, click on **Fields**. Read all the messages, and under the **Entity and Attribute Information**, click on the **Details: WorldCountries** arrow to expand it. Click on **Entity Type**, and for **Definition**, type Admin 0-Countries. For **Definition Source**, type Natural Earth.

18. Click on **Entity Type**, and for **Definition**, type Admin 0-Countries. For **Definition Source**, type Natural Earth.

19. To resolve the rest of the messages, we should correct only the user-defined fields: type, name, pop_est, gdp_md_est, and continent, as shown in the following screenshot:

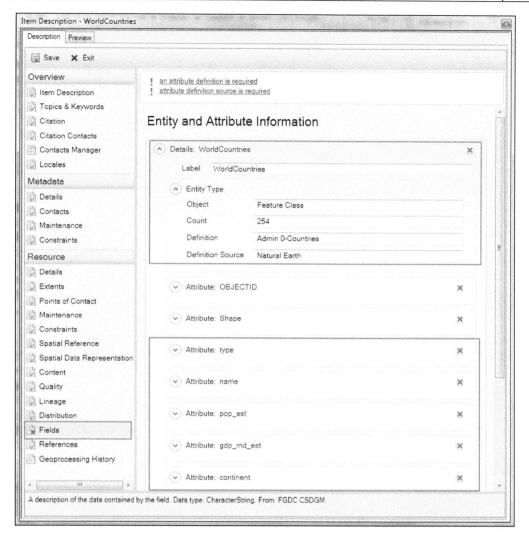

20. Click on **Attribute: type**, and for **Definition**, type `Country type`. For **Definition Source**, type `Natural Earth`. To fix the `at least one type of attribute domain is required` message, click on the **Unrepresentable Domain** button, and type `Country types stored as String`.

21. Repeat the last step for the rest of the user-defined fields. After you successfully complete the metadata of the `WorldCountries` feature class, save the changes and exit the metadata editor by clicking on the **Save** button. Inspect the metadata in the **Description** panel. Close the **Item Description** window.

# Importing metadata

As you can see, documenting your datasets is a time-consuming process, and sometimes, this could become an expensive stage in the GIS data production workflow. In this section, we will continue to document our feature classes.

[  Please perform the steps mentioned in this section in a single session. ]

As the data source of the WorldNationalCapitals feature class is also made by Natural World, we will create metadata for it by importing the information from the WorldCountries feature class. Importing metadata information from one feature class to another will save us a lot of time and effort:

1. In the **Catalog** window, right-click on the **WorldStatesProvinces** standalone feature class and select the **Item Description**. Create a thumbnail and inspect the attribute table in the **Preview** section.

2. Return to the **Description** section, click on the **Import** button, and set the parameters, as shown in the following screenshot:

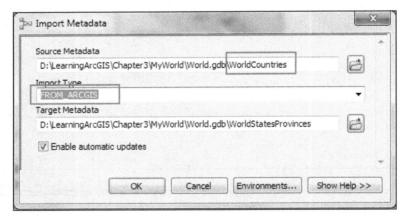

3. Click on **OK** and then click on **Close**. To see the updated metadata content, close the **Item Description** window and open it again. Click on **Edit** to open the metadata editor. Most of the metadata elements have been populated, but you should adjust some of them slightly, such as **Tags**, **Description**, and **Fields**.

In the next steps, we will document the `EuropeCities_3m` feature classes from the `Europe` feature dataset using the existing metadata files in the XML format distributed by the Eurostat:

4. In the **Catalog** window, navigate to `..\MyWorld\Eurostat\URAU_2001\ metadata`. Right-click on the **Urban_Audit_2001.xml** metadata file and click on **Item Description**.

5. Explore the metadata elements from the ISO19139 metadata format. Keep the **Item Description** open.

6. In the **Catalog** window, select the `EuropeCities_3m` feature class to display the metadata of the feature class in the **Item Description** window. Create a thumbnail and inspect the attribute table in **Preview** section.

7. Return to the **Description** section, click on the **Import** button, and set the parameters, as shown in the following screenshot:

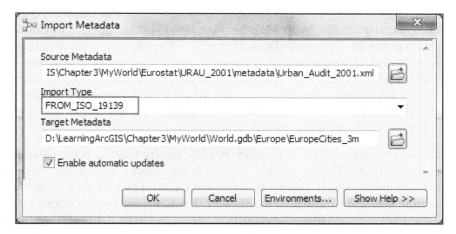

8. Click on **OK** and then click on **Close**. Close the **Item Description** window and open it again to update the metadata content. Click on **Edit** to open the metadata editor.

> Even if the XML metadata file format has the XML schema in compliance with the EN ISO 19139 standard schema, this doesn't mean that the XML metadata file will contain information for all mandatory elements of the metadata profile. You should always check whether all mandatory elements defined by a standard are completed.

9. In the **Resource** section, there are three incomplete elements of metadata: **Quality**, **Distribution**, and **Fields**. Click on **Quality**, and for **Scope Level**, choose **Dataset**. Click on **Distribution** and expand the **Distribution Format** arrow. In the **Format Version** field, type 10.4. As your dataset is only for educational proposes and the linkage URLs are not mandatory in the ISO19115 metadata standard, remove all four **Digital Transfer Options** by clicking on the red **X** on the left.

> The ISO 19139 XML schema implementation is derived from the ISO 19115 Geographic information-Metadata standard. For more details, please refer to:
>
> http://www.iso.org/iso/catalogue_detail.htm?csnumber=32557

10. Try to repair the **Fields** metadata element that is marked with a red **X**.

    The Urban_Audit_2001.xml metadata file downloaded from Eurostat refers to an available **Web Map Service (WMS)** that allows a minimum interaction with viewable datasets, such as display geometry and legend information, navigate, overlay, zoom in, zoom out, or pan. An example of a public view service is the Eurostat Statistical e-Atlas at http://ec.europa.eu/eurostat/statistical-atlas/gis/viewer/. As our geodatabase is a local dataset, we will continue to adjust the metadata information of the EuropeCities_3m feature class.

11. Under **Metadata**, click on **Details**. For **Hierarchy Level**, choose **Dataset**. Note that under **Overview**, the **Topics & Keywords** option has been marked as incomplete. Select **Topics & Keywords**, Under the **Topic Categories** section, select **Boundaries and Society**. For **Content Type**, choose **Offline Data**. Expand the **Other Keywords** element to see its content. Remove this element by clicking on the red **X** on the left. Click on **OK**. Click on **Save** and inspect the metadata.

12. Let's export the `EuropeCities_3m` feature class metadata. Click on the **Export** tool and set the parameters, as shown in the following screenshot:

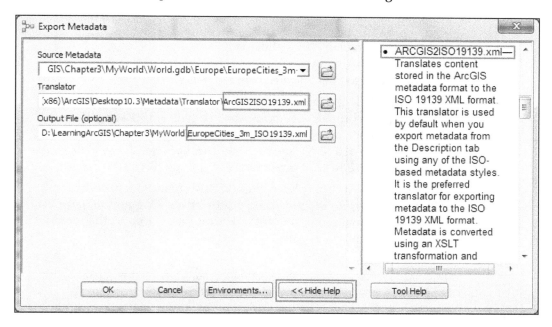

13. Click on the **Help** button and click on the **Translator** field to read more information about the default translator, `ArcGIS2ISO19139.xml`. Click on **OK** to run the tool.

14. You can inspect the resulting XML metadata file using the **Item Description** window from the **Catalog** window, using an internet browser, or better yet using a free code editor such as *Notepad++*.

Finally, we will change the metadata style in ArcMap to the INSPIRE style, which is the European metadata standard:

15. From the **Customize** menu, select **ArcMap Options** and click on the **Metadata** tab. For **Metadata Style**, select the **INSPIRE Metadata Directive** style from the drop-down list. Click on **Apply** and then click on **OK**.

16. In the **Catalog** window, right-click on the **EuropeCities_3m** feature class and select **Item Description**. Click on **Edit** to inspect the metadata elements. As you can see the `INSPIRE` style has different mandatory elements, such as **Constrains**, **Lineage**, and at least one linkage URL for the dataset for the **Distribution** element.

We recommend that you download the technical guide of the INSPIRE metadata at http://inspire.ec.europa.eu/documents/Metadata/MD_IR_and_ISO_20131029.pdf. On page 10, you can see the metadata differences between the ISO 19115 and INSPIRE standards.

17. Try to repair the **Constrains** and **Lineage** elements.

18. For **Distribution,** click on **New Digital Transfer Options**. For **Units of Distribution,** type Personal geodatabase. To fix the **linkage URL is required** message, click on **New Online Resource** and type the following URL: http://ec.europa.eu/eurostat/web/gisco/geodata/reference-data/administrative-units-statistical-units. Press the *Tab* key.

19. After you've successfully completed the INSPIRE mandatory elements, save the changes, and exit the metadata editor by clicking on the **Save** button. Inspect the metadata in the **Description** panel. Export the metadata information as EuropeCities_3m_INSPIRE.xml.

It's good practice to create metadata for all levels of your geodatabase. For instance, create metadata for all your feature classes stored in the file geodatabase. Then, create metadata for the feature datasets existing in your geodatabase. Use the **Import** tool to add common metadata information from the feature class level, such as **Metadata** | **Contacts**, **Resource** | **Points of Contact**, and **Constraints**.

Lastly, right-click on the **World.gdb** file geodatabase and open **Item Documentation** to edit metadata at the geodatabase level.

20. Close **Item Description** and exit ArcMap without saving the map document.

You can find the results of this exercise at <drive>:\LearningArcGIS\Chapter3\Results\+Metadata.

# Summary

In this chapter, you built a file geodatabase schema while working with three primary components of a geodatabase: feature classes, feature datasets, and nonspatial tables. In addition to this, you explored different ways to add initial data to your file geodatabase. You also learned that spatial reference is an important property of the standalone feature classes and feature datasets.

In the second part of this chapter, you learned how to document your geodatabase using two metadata standards: the international standard, ISO19139, and the European standard INSPIRE.

In the next chapter, you will learn how to create map symbology for our `World.gdb` file geodatabase using symbols and labels in the ArcMap application.

# 4
# Creating Map Symbology

In this chapter, we will start working with symbols and labels. Based on the characteristics of our data, we will create custom symbols to represent the geographic objects on different thematic maps. Moreover, we will learn how to map the density of the world's population by symbolizing a map layer based on its qualitative and quantitative attribute values.

By the end of this chapter, you will learn about the following topics:

- Creating custom point, line, and area symbols within map layers
- Labeling map features
- Working with the layer files
- Mapping qualitative and quantitative data

## Creating symbols

ArcGIS for Desktop allows you to display geographic features using symbols on the map based on their attributes. With symbols, you can display unique values or ranges of values on a map. The resulting maps are also known as thematic maps. There are two types of thematic maps: qualitative and quantitative. Thematic maps can represent four types of attribute data, as follows:

- **Nominal** (qualitative maps): This refers to the category or discrete values (name of countries, soil types, or land use categories)
- **Ordinal** (quantitative maps): This refers to the category values that are ordered based on rank or importance (classification of rivers, levels of education, or vehicle traffic)
- **Interval** (quantitative maps): This refers to the quantity values that compare the relative importance of values having an arbitrary zero (temperature, scale, or pH values)

- **Ratio** (quantitative maps): This refers to the quantity values that compare the relative importance of values having an absolute zero (population, height, or age)

Symbols have graphic characteristics, which can vary in order to emphasize the differences, hierarchy, or importance of features on a map.

The graphic characteristics or visual variables refer to symbol size, color value (gray tone), hue (color), texture (pattern), orientation, and shape. These characteristics can be applied individually or in combination to the point, line, and area symbols, depending on the type of attribute data that has to be represented on a map.

The following table shows how visual variables of the symbols should vary on thematic maps:

| Graphic characteristics of symbols | Qualitative thematic map | Quantitative thematic map |
|---|---|---|
| Size | - | ✓ |
| Lightness (Color value) | - | ✓ |
| Color (Hue) | ✓ | - |
| Arrangement (line and area only) | ✓ | - |
| Separation (line and area only) | - | ✓ |
| Orientation (angle) | ✓ | - |
| Shape | ✓ | - |

Karen K. De Valois and Russell L. De Valois have mentioned that color has three perceptual dimensions: **Hue**, **Lightness**, and **Saturation** (Color Vision, Elsevier Academic Press, pages 129-175). Hue is defined by mixing the following color of light or *additive primaries*: Red, Green, and Blue (**RGB**). RGB is the main color system used by computer screens and televisions.

Hue can be also defined by mixing the following colors or *subtractive primaries*: Cyan, Magenta, and Yellow (**CMY**). **CMYK** is the main color system used for desktop printing and commercial offset printing. CMYK refers to four basic ink colors: Cyan, Magenta, Yellow, and Black (K stands for key).

[  Here is a nice story of why the letter K represents black at
https://gearside.com/color-black-represented-k-cmyk/. ]

The Value (lightness or darkness) is the quantitative dimension of a color and defines how light or dark the color actually is. The Saturation (intensity) is a third qualitative dimension of the color and defines how bright or dim the color is.

In ArcMap, you can mix the color in RGB and CMYK color systems, or through another three-dimension color space called **HSV** that links the Hue, the Saturation, and the Lightness (or Value).

# Creating point symbols

A point symbol can vary in size, shape, color (hue and lightness), and angle. A point symbol size can vary proportionally (proportional to the absolute or relative data value that it represents) or gradually (graduated to the number of classes). You can use the symbol angle when you want to vary a point symbol that has a constant size and shape in the map legend.

In this section, we will learn how to save a map layer as a standalone layer file (.lyr). A layer file saves all the properties of a layer, including symbology, labels, source, and more.

 The steps in this section have been broken down into multiple exercises. However, they all need to be performed in one single session.

Follow these steps to start working with point symbols using the ArcMap application:

1.  Start the ArcMap application to open an existing map document, Symbols. mxd, from <drive>:\LearningArcGIS\Chapter4\Symbols. The data frame's coordinate system is WGS 1984 World Mercator. You have already worked with the *Mercator* projection in *Chapter 2, Using Geographic Principles*.

2.  From the **Bookmarks** menu, select **Europe** to take a closer look at our map. In **Table Of Contents**, right-click on the **PopulatedPlaces** layer and select **Open Attribute Table**. Scroll through the table to inspect the attributes associated with features in the PopulatedPlaces layer. We will use the FEATURECLA attribute field that classifies the cities to create point symbols on a qualitative thematic map. Close the **Table** window.

 To preview the results of this exercise, open the PDF file named PointSymbols_1.pdf, which is stored at <drive>:\LearningArcGIS\Chapter4\Symbols.

3.  In **Table Of Contents**, turn off the `CostlineBuffer` layer. Right-click on the **PopulatedPlaces** layer, select **Properties** and click on the **Symbology** tab. All cities are currently using the same point symbol.

We will now differentiate the capital from the rest of cities by keeping the symbol size constant and varying the symbol color, as shown in the following screenshot:

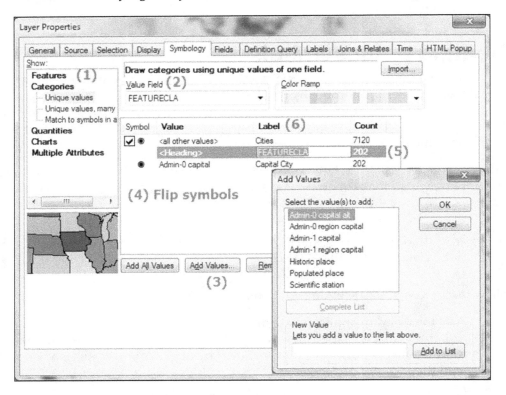

4.  As seen from the screenshot, in the **Show** area, click on **Categories**. **Unique values** should be selected. From the **Value Field** drop-down list, select **FEATURECLA**. Click on the **Add Values** button. To see all unique attribute values, click on **Complete List**. Select **Admin-0 capital** and click on **OK** to add it to the symbol list.

5.  In the **Layer Properties** dialog window, click on the **Symbol** column heading and select **Properties for All Symbols** to change both symbols at once. In the **Symbol Selector** window, scroll down the list of default symbols and click the **Circle 3** symbol. For **Color**, choose **Grey 30%**. Change **Size** to 10 points. Click on the **Edit Symbol** button to modify the symbol.

6.  In the **Symbol Property Editor** window, select the first symbol layer in the **Layers** section. In the **Properties** section, note the type of symbol from the first symbol layer, **Character Marker Symbol**. For **Unicode**, type 80 to select a new character marker for the symbol layer, as shown in the following screenshot:

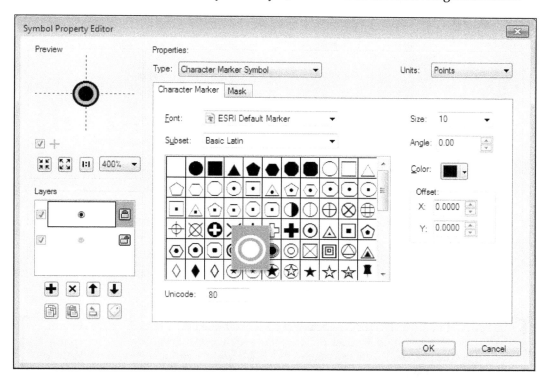

7.  Click on **OK** twice to return to the **Layer Properties** window. Click on **Apply** to see your changes on the map.

8.  We will change the symbol color for the Admin-0 capital features to reinforce the categorical differences between ordinary cities and capitals. Double-click on its symbol patch to open the **Symbol Selector** window and click on the **Edit Symbol** button. Select the background layer (second layer), and change its **Color** option to red. Click on **OK**.

9. In the **Symbol Selector** window, click on the **Save As** button and type the name and category of your new symbol, as shown in the following screenshot:

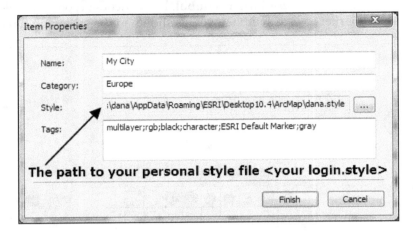

10. Click on **Finish** to save the symbol in your default personal style file stored at `<drive>:\Users\<your login>\AppData\Roaming\ESRI\Desktop 10.4\ArcMap\<your login.style>`.

11. In the **Symbol Selector** list, you should see your new symbol. Click on **OK** to return to the **Layer Properties** window.

If you want to modify or remove a customized symbol from your default personal style, from the **Customize** menu, select **Style Manager**. Expand the personal style set and select the **Marker Symbols** folder to see your point symbol. Right-click on the symbol, select **Delete** to erase the symbol, or select **Properties** to change the symbol.

12. Next, we will remove the heading label called FEATURECLA, which is displayed in **Table Of Contents**. Under the **Label** column, select the legend's heading title, <Heading>. Click on the FEATURECLA title and remove it.

13. Change the <all other values> labels to Cities and Admin-0 capital to Capital Cities. Click on **Apply**.

14. Next, click on the **General** tab and change **Layer Name** to PopulatedPlaces_Capitals. Click on **OK**.

As we cannot see the differences between the cities and capitals on the map display at the scale 1:25 million and the reference scale for source data is 1:10 million, we will define a visible scale range for the PopulatedPlaces_Capitals layer. In *Chapter 2, Using Geographic Principles*, you learned how to set the minimum and maximum visible scale for a layer.

15. Set **Minimum Scale** to 10 million and **Maximum Scale** to 3,000,001. Type 10,000,000 scale in the map scale list and press the *Enter* key. In **Table Of Contents**, right-click on the **PopulatedPlaces_Capitals** layer and navigate to **Visible Scale Range | Set Minimum Scale**. Repeat this step, to set **Maximum Scale**.

16. If you want to see the names of the cities while you are inspecting the map using the **Zoom in, Zoom out,** and **Pan** tools, navigate to **Layer Properties | Display**. For **Display Expression**, select the **NAME** field from the drop-down list and select **Show MapTips using the display expression**. Close the **Layer Properties** window. On the map, pause the pointer on a point feature to see the city's name.

In the next exercise, we will create a copy of the PopulatedPlaces_Capitals layer so that we can apply it to graduated point symbols based on quantitative data, which is stored in the POP_MAX attribute field. Follow these steps to start to symbolize the layer:

1. In **Table Of Contents**, right-click on the **PopulatedPlaces_Capitals** layer and select **Copy**. Right-click on the **World** data frame, and select **Paste Layers (s)**. The duplicated layer is added to the top of **Table Of Contents**. Change the name of the duplicate layer to PopulatedPlaces_Population (select it and click on the layer again to make it editable). Turn off the PopulatedPlaces_ Capitals layer.

   This step will not duplicate the PopulatedPlaces feature class in our file geodatabase. Both layers display the same feature class called PopulatedPlaces from <drive>:\LearningArcGIS\Chapter4\World.gdb geodatabase. Try to verify this by looking at the source of each layer.

2. From the **Bookmarks** menu, select Europe.

 To preview the results of this exercise, open the PDF file named PointSymbols_2.pdf, which is stored at <drive>:\LearningArcGIS\Chapter4\Symbols.

3. Double-click on the PopulatedPlaces_Population layer to open **Layer Properties**. Select the **General** tab, check **Show layer at all scales**, and then click on **Apply**. Select the **Symbology** tab. In the **Show** area, click on **Quantities**, and then click on **Graduated symbols**.

Next, we will create four population range classes with point symbols having the same color and graduated sizes, as shown the following screenshot:

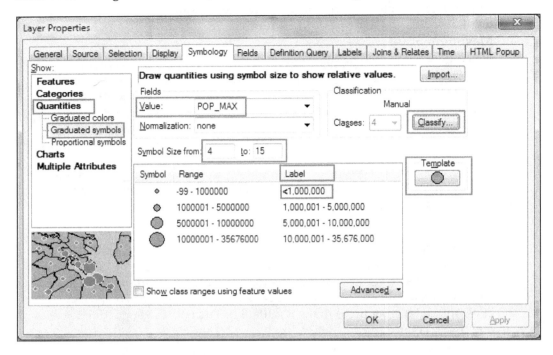

4.  In the **Fields** tab, from the **Value** drop-down list, select POP_MAX. By default, ArcMap has classified population values into five classes using the `Natural Breaks (Jenks)` classification scheme. Click on the **Classify** button.

    Under **Classification Statistics**, inspect the statistical parameters, such as `Mean`, `Median`, or `Standard Deviation`. The classification histogram represents the number of countries (vertical axis) for each unique population value (horizontal axis) that is stored in the POP_MAX field. The height of every bar indicates the frequency of population values. You can zoom in using the histogram's context menu (right-click on it).

    The histogram has one prominent peak (unimodal) and the distribution of population values is skewed to the right. Even if the `Natural Breaks (Jenks)` classification scheme is suited for values which are not evenly distributed, we would like to break the classes at specific values and keep the number of classes as low as possible.

5.  Select 4 from the **Classes** drop-down list, and then select **Manual** for the **Method** option in the **Classification** section.

6. We want to exclude all cities with a population less than 100,000. Click on the **Exclusion** button and build the `POP_MAX < 100000` expression. Click on **OK** to apply the exclusion and close the query builder window.

   Inspect the histogram of the data values. The vertical axis represents the number of cities and horizontal axis represents the population values. Notice the breaks values displayed with blue lines.

7. You can manually change the break values by dragging the blue lines or by typing the values in the **Break Values** list on the right, as shown in the following screenshot:

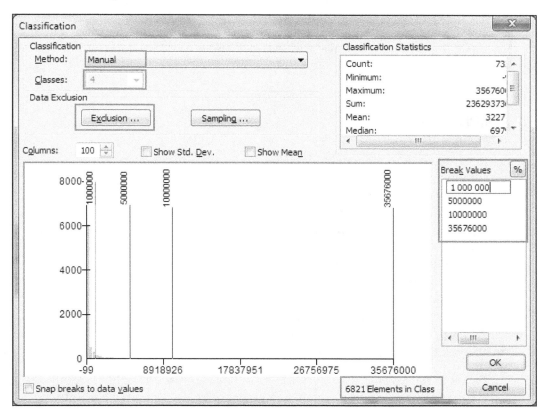

8. Edit only the first three values (1,000,000; 5,000,000; 10,000,000) and leave the last value that represents the maximum values that is stored in the `POP_MAX` field unchanged. Click on **OK** to close the **Classification** dialog window.

9. Set **Symbol Size** from 5 to 25. Click on the **Label** column heading and select **Format Labels**. In the **Number Format** window, select **Show thousands separators** and click on **OK**. Change the label of the first class from -99-1,000,000 to <1,000,000. Change the color of the symbols to **Tarragon Green** by clicking on the **Template** button on the right-hand side.

10. Click on **Color** and choose **More Colors**. The **RGB** color model should be selected. For **Red**, **Green**, and **Blue**, type the following values: 112, 168, 0.

Even if the symbol sizes change gradually, they have no connection to the actual population values. If you want to create a connection between symbol size and population value, use **Proportional symbols**. However, it is not recommended to use proportional symbols when there is a large range of values because the symbols will overload the map by their number and size.

Click on **OK** until all dialog windows are closed.

11. In **Table Of Contents**, select the legend header text, POP_MAX. Click on it again and change the text to Population.

12. Refresh the map display by pressing the *F5* key and inspect the point symbols at different map scales using the **Zoom in** and **Zoom out** tools. Note that the size of the symbols does not change as the map scale changes.

13. As the reference scale of source data from *Natural Earth* is 1:10 million, we will set the same reference scale for point symbols on the map. On the **Standard** toolbar, for **Scale**, type 10,000,000 and press the *Enter* key. In **Table Of Contents**, right-click on the **World** data frame and navigate to **Reference Scale | Set Reference Scale**.

Now, with a reference scale set, our point symbols will scale with the map display. Use the **Zoom out** tool to see how symbols get smaller and use **Zoom in** to see how the symbols get larger. The symbol maintains its size relative to the data frame reference scale.

14. To zoom the layer at the reference scale that you just set, right-click on the **World** data frame and navigate to **Reference Scale | Zoom To Reference Scale**.

We can see the differences between four population classes on the map displayed at scale 1:20 million. Therefore, we should extend the visible scale range of the PopulatedPlaces_Population layer as follows: minimum scale to 1:20 million and maximum scale to 1: 3,000,001.

In the next steps, you will symbolize the Ports layer using a pictogram symbol from the ArcMap collection of symbols, which comes with over 12,000 symbols, which are stored in different style files. Each style file generally refers to a specific industry or field of practice:

15. In **Table Of Contents**, double-click on the symbol patch below the Ports layer. In **Symbol Selector**, click on **Style References**, select the Military METOC style, and then click on **OK**.

16. In the **Symbol Selector** window, select **Referenced Style**, type anchor, and click on the small search button on the right-hand side. Select the **Anchorage Point** symbol. Click on the **Edit Symbol** button and note that the symbol type is **Picture Marker Symbol** and has only one layer. Change **Size** to 8 and click on **OK** three times to close the dialog windows. Inspect the results.

In the last part of the exercise, we will save the PopulatedPlaces_Capitals and PopulatedPlaces_Population layers as two separate layer files (.lyr).

You may wonder, why should I start working with layer files? For example, you may symbolize a map layer for different thematic maps or for different map scales without duplicating the source data. A layer file helps you standardize symbology throughout your organization. You can easily share only the layer files that already have access to source data with your colleagues.

Follow these steps to start creating layer files:

1. In **Table Of Contents**, right-click on **PopulatedPlaces_Capitals** and click on **Save As Layer File**. Navigate to the <drive>:\LearningArcGIS\Chapter4\ Symbols folder and retain the default **Name** of the layer.

 From the **Save as** drop-down list you can choose to save the layer to a previous version of ArcGIS for Desktop.

2. Next, click on **Save**. Repeat the previous step to save PopulatedPlaces_ Capitals as a layer file.

3. Then, open the **Catalog** window, connect to the <drive>:\LearningArcGIS\ Chapter4 folder, and expand the **Symbols** folder. You should see the layer files that you just created.

4. Let's test the functionality of these layer files. Remove these two layers from **Table Of Contents**; and from the **Bookmarks** menu, select the **Europe** bookmark. Drag and drop the layers from the **Catalog** window into the map display area. Note that the layers have kept their symbology.

5. In **Table Of Contents**, review **Layer Properties**, including the **Source** and **Symbology** settings, to see that they are the same as the layers that you removed.

6. When finished, save your changes to the map document. From the **File** menu, select **Save As**, navigate to the `..\LearningArcGIS\Chapter4\Symbols` folder, and save your map as `MySymbols.mxd`. Leave the map document open to continue working with line symbols in the next subsection.

You can find the results of this exercise at `<drive>:\LearningArcGIS\Chapter4\Results\Symbols\PointSymbols.mxd`.

# Creating line and area symbols

A line symbol can vary in width (or size), color (hue and lightness), and pattern (dashed or cased lines). A line symbol width can vary proportionally (representing the absolute or relative data values) or gradually (is not proportioned to data values).

The dashed line pattern can combine visual variables, such as shape, angle, and separation. A cased line pattern works with at least two symbol layers, which combine all visual variables mentioned previously.

The area patterns can combine visual variables, such as hue, lightness, shape, angle, separation, and arrangement (grid or random).

In this section, we will use another method to exclude features from the map display named **Definition Query**. A definition query allows you to display only a subset of features from a layer for cartographic purposes.

Follow these steps to start working with line and area symbols using the ArcMap application:

1. Open your map document called `MySymbols.mxd`. In **Table Of Contents**, collapse and turn off the `PopulatedPlaces_Capitals`, `PopulatedPlaces_Population`, and `Ports` layers. Click on the – icon next to the layer to collapse it. Deselect the small checkbox next to the layer to turn it off.

 To preview the results of this exercise, open the PDF file named `LineAreaSymbols.pdf`, which is stored at `<drive>:\LearningArcGIS\Chapter4\Symbols`.

2. Turn on the `Roads` layer, open its attribute table, and scroll through the table to inspect the attributes associated with features in the `Roads` layer. The `type` and `continents` fields will help us display and classify features on the map. Close the attribute table.

3. We would like to display only the road features from the European continent on the map without affecting the source data. In **Table Of Contents**, double-click on the **Roads** to open the **Layer Properties** window.

4. Click on the **Definition Query** tab and click again on the **Query Builder** button. Build the `continent='Europe'` expression. Click on **OK**.

5. In the **Layer Properties** window, click on **Apply**. Then, select the **Symbology** tab. In the **Show** area, click on **Categories**. **Unique values** should be selected. Under the **Value Field**, select **type** from the drop-down list.

6. Click on the **Add Values** button and click on **Complete List**. Select **Major Highway** and **Secondary Highway** using the *Ctrl* key. Click on **OK**.

7. Let's adjust the default label and symbol colors. Change the legend's heading title to **Highway** and the feature labels to **Major** and **Secondary**, as shown in the following screenshot:

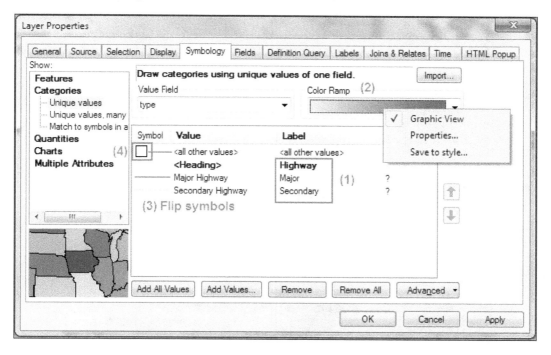

8. Right-click on **Color Ramp** and deselect the **Graphic View** to display the color ramp names. Select the color ramp called **Errors**. Right-click on **Color Ramp** again and select **Graphic View**.

9. Next, we will set a darker color for the `Major` highway and a lighter color for the `Secondary` highway. Click on the **Symbol** column heading and select **Flip Symbols**.

10. Click on the **Symbol** column heading again and select **Properties for all symbols**. Change **Width** to 0.5. Click on **OK**.

11. Deselect the <all other values> checkbox. Click on **Apply** to verify your changes in **Table of Contents** and on the map. When finished, click on **OK**. Inspect the results using different map scales. When finished, save your changes to the map document by clicking on the **Save** tool from the **Standard** toolbar.

In the next steps, we will symbolize the RiversLake_Centerlines layer:

12. Turn off the Roads layer. Turn on the RiversLake_Centerlines layer and open its **Layer Properties** window. The **Symbology** tab should be selected. In the **Show** area, navigate to **Categories | Unique values**. For the **Value Field**, select **featurecla**.

13. Click on the **Add All Values** button. We have two types of features: Lake Centerlines, and Rivers.

14. We will change the order of two types of features in the legend. Click on the **Value** column heading and select the **Reverse Sorting** option. Click on **Apply**.

15. In this step, we will customize the two line symbols. Double-click on the line symbol for River to open the **Symbol Selector** window. To change the color, click on the color square and select **More Colors**. In the **Color Selector** window, for **Red**, **Green**, and **Blue**, type the following values: 115, 178, 255, as shown in the following screenshot:

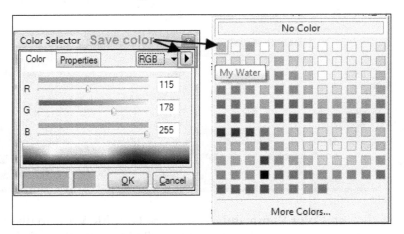

16. Click on the black arrow at the top-right corner to save the color as My Water. Click on **OK** twice to return to the **Symbol Selector** window. Change **Width** to 0.5. Save the customized line symbol as My River with **Category**: Water. Click on **OK** to close the window.

In your default personal style set, the **My Water** color is stored in the `Color` folder and the line symbol called `My River` is stored in the `Fill Symbols` folder. The custom colors and symbols saved to your personal style will not be available to others.

17. If you want to indicate river flow (based on the direction of the polyline) select the **Edit Symbol** button in the **Symbol Selector** window. Change **Type** to **Cartographic Line Symbol**. Click on the **Line Properties** tab and for **Line Decorations**, select the second arrow. To change the color and size of arrow, use the **Properties** button.

18. Double-click on the line symbol for `Lake Centerline` to open the **Symbol Selector** window. Click on **Edit Symbol** to open the **Symbol Property Editor** window. Change the color to **My Water**. In the color palette window, you should see your color called **My Water** at the top-left corner.

19. Select **Cartographic Line Symbol** from the **Type** drop-down list. Select the **Template** tab to create your own dashed line pattern, as shown in the following screenshot:

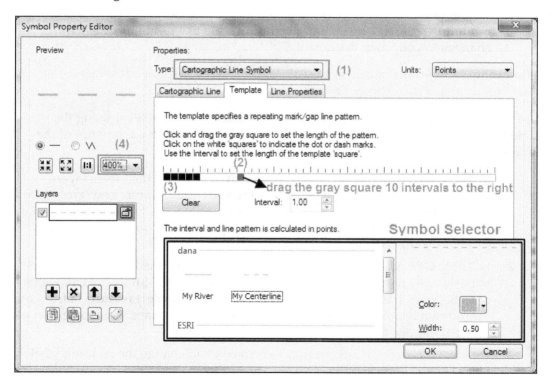

20. Drag the small gray square 10 intervals to the right to create a pattern formed by 5 black and 5 white squares. Then, click on the first five white squares to turn them black. In the **Preview** section, change the zoom to 400% to see the dashed pattern. Click on **OK**.

21. In the **Symbol Selector** window, save the new dashed line symbol as My Centerline, as shown in the previous screenshot. Click on **OK**.

22. In the **Layer Properties** window, delete the <Heading> label text and deselect <all other values>. Click on **OK**.

23. To zoom the layer at the reference scale that you just set, right-click on the **World** data frame and navigate to **Reference Scale | Zoom To Reference Scale**. Remember that the World data frame has the reference scale defined at 1:10 million. Inspect the line symbols at different scales. Note that line symbols will scale with the map display.

24. Right-click on the **World** data frame and navigate to **Reference Scale | Clear Reference Scale**. Again, display the line symbols at different scales. Your symbols are not scaled with the map display anymore.

25. Set the visible scale range of the RiversLake_centerlines layer as follows: minimum scale to <None> and maximum scale to 1:3 million. Save the changes to your map document.

You can find the results of this section at <drive>:\LearningArcGIS\Chapter4\ Results\Symbols\LineSymbols.mxd.

In the second part of this exercise, we will symbolize a polygon layer using the **Categories** classification scheme. We will vary the following characteristics of the area symbols: color, texture, and orientation.

1. In **Table Of Contents**, turn off the Road layer. Turn on the Lakes layer. Right-click on the **Lakes** layer, select **Open Attribute Table**, and scroll through the table to inspect the attributes associated with features in the Lakes layer. The featurecla field stores nominal (qualitative) data. Close the **Table** window.

2. In **Table Of Contents**, double-click on **Lakes** to open the **Layer Properties** window. The **Symbology** tab should be selected. In the **Show** area, navigate to **Categories | Unique values**. Under **Value Field**, select **featurecla** from the drop-down list. Add all values in the legend list. There are three categories of lakes: Alkaline Lake, Lake, and Reservoir.

3. Deselect the <all other values> checkbox and change the heading label text to Type.

4. Next, we will change the default solid fill symbols that were assigned to all three categories. Double-click on the symbol patch next to the `Lake` layer and click on the **Edit Symbol** button. In the **Symbol Property Editor** window, for **Color**, add the `Sodalite Blue` color RGB: `190,232,255`. For the outline symbol, we will reuse a line symbol created in the first part of this exercise. Click on the **Outline** button, and select the **My River** line symbol. Click on **OK** twice to return to the **Symbol Selector** window. Save this area symbol as `My Lake`. Click on **OK**.

5. In this step, we will create a hatch pattern for alkaline lakes. Double-click on the symbol patch next to the `Alkaline Lake` and click on **Edit Symbol**. From the **Type** drop-down list, select **Line Fill Symbol**. Change **Color** to **My Water**. Click on the **Outline** button, and select the **My River** line symbol. Click on **OK**. Change **Angle** to `-45` and **Separation** to `2`. Click on **OK**. Save this area symbol as `My Alkaline Lake`. Click on **OK** to close the **Symbol Selector** window.

6. Double-click on the symbol patch of `Reservoir`. Scroll down the list of the symbol and click on the `Reservoir` symbol. You may change the symbol properties and save it in your personal style set. In your default personal style set, the area symbols that you have created are stored in the `Fill Symbols` folder. Click on **Apply**. The new symbols for lakes displays in **Table Of Contents** and on the map.

   We will draw the lakes over the rivers, but we also want to see the lake centerline, as shown in the following screenshot:

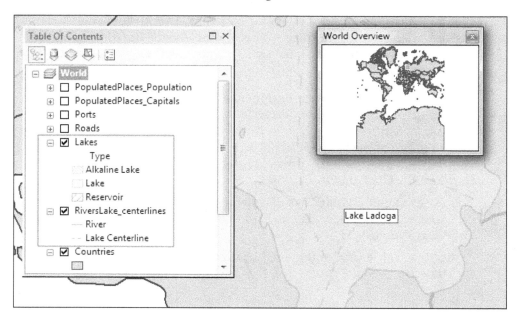

7. Select the **Display** tab. Change the **Transparent** value to 50%, select the **name** field from the **Display Expression** drop-down list, and select **Show MapTips using the display expression**. Click on **OK** to save changes and close the **Layer Properties** window.

8. In **Table Of Contents**, select **Lakes** and drag it above the **RiverLake_ centerlines** layer. From the **Bookmarks** menu, select **Lake** Ladoga (Russia) to see the lake from the previous screenshot. Inspect the results. Reordering the layers in **Table Of Contents** will change the order of how features are drawn on the map.

9. Set the visible scale range of the Lakes layer as follows: minimum scale 1:20 million and maximum scale 1: 3 million. Save the changes in your map document.

In the last part of this exercise, we will use a layer file from the European Environment Agency (EEA) website to symbolize the Europe_Cities10k feature class from the Europe feature dataset:

1. From http://www.eea.europa.eu/data-and-maps/data/urban-atlas/, click on the **Additional information** tab and download the **Esri ArcMAP(Octet Stream)** file to ..\Chapter4\Symbols.

2. In the **Catalog** window, click on **UrbanAtlas_rgb.lyr** and drag it above the **Countries** layer. The red exclamation mark tells you that the UrbanAtlas layer can't find the source data. Double-click on the layer to open the **Layer Properties** window. Click on the **Source** tab and inspect the source details.

   We don't have the n10031 shapefile stored at J:\data\UrbanAtlas. Instead, we have the two cities—Paris and Brussels—stored in a single feature class with the same table structure and attribute values as the n10031 source shapefile.

3. Click on the **Set Data Source** button, navigate to <drive>:\ LearningArcGIS\Chapter4\World.gdb\Europe, and select the Europe_Cities10k feature class. Click on **Add**. Note that the coordinate reference system (CRS) of the feature class is different from the data frame's CRS. ArcMap will automatically project on-the-fly the data to match the data frame's CRS. Click on **OK**.

4. The red exclamation mark has disappeared. In **Table of Contents**, right-click on UrbanAtlas and select **Zoom To Layer**. You should see the two cities, Paris and Brussels.

5. Notice that **Table Of Contents** has changed to **List By Source**. Click on **List By Drawing Order**.

Next, we will display only the fast transit roads and water from Brussels at scales, 3,000,000 to 5,000:

6. Open the **Layer Properties** window again, and select the **General** tab. Change the UrbanAtlas layer's name to EuropeCities_10k. Select **Don't show layer when zoomed** and set the minimum scale to 3,000,001 and the maximum scale to 5,000.

7. Click on the **Definition Query** tab and click on the **Query Builder** button again. Build the CITIES = 'Bruxelles/Brussel' AND CODE = '12210' OR CITIES = 'Bruxelles/Brussel' AND CODE = '50000' expression. Click on **OK**.

8. In **Table of Contents**, right-click on **EuropeCities_10k** and select **Zoom To Layer**. Inspect the results.

9. From the **Bookmarks** menu, select **Europe**. Turn on the following layers: PopulatedPlaces_Population, Roads, and CoastlineBuffer.

Let's change the data frame's background color, as shown in the following screenshot:

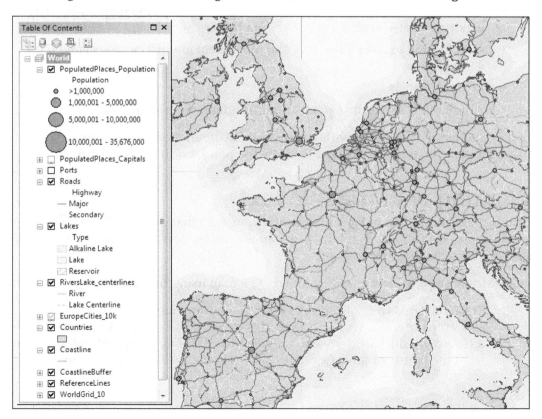

10. Double-click on the **World** data frame to open the **Data Frame Properties** window. Click on the **Frame** tab and in the **Background** area, click on the down arrow to select **Blue**. Click on **Apply** to see the map updates.

11. To change **Color**, click on the color square and select **More Colors**. For **Red**, **Green**, and **Blue**, type the following values: 230, 250, 255. Save the color as My Data Frame. Inspect the results.

12. When finished, save the changes to your map document called MySymbols. mxd. Close ArcMap.

You can find the results at <drive>:\LearningArcGIS\Chapter4\ Results\ Symbols\AreaSymbols.mxd. All resulting layer files are stored at:..\Chapter4\ Results\Layers.

# Creating labels

A label describes a feature through map text. There are four types of map text:

- Graphic text in **Layout View**, such as title or notes on the map layout that is not associated with a particular data frame

- Graphic text in **Data View** or a map document annotation associated with a data frame

- Dynamic labels

- Annotations stored in a geodatabase as annotations feature class

In this chapter, we will work with map document annotations and dynamic labels using the same layers that we used to create the point, line, and area symbols.

## Working with graphic text

Graphic text allows you to label features on your map that do not exist in your geodatabase. Text added manually to a map is called a map annotation and is stored within the map document.

Follow these steps to start adding map text to a map using the ArcMap application:

1. Start the ArcMap application to open an existing map document, Labels. mxd, from <drive>:\LearningArcGIS\Chapter4\Labels. As we don't have a layer with the world oceans, we will add a label text manually on a map to label the North Atlantic Ocean, as shown in the following screenshot:

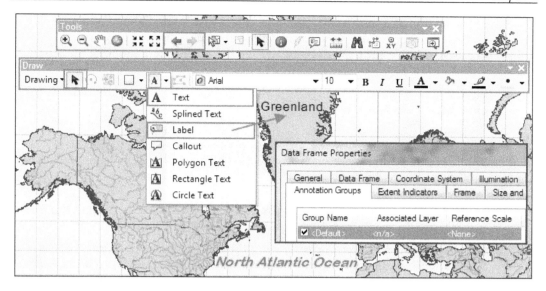

2. From the **Customize** menu, select **Toolbars** and select **Draw**. On the **Draw** toolbar, click on the **New Text** button. Click anywhere between *United States* and *Europe*. The map text is too small to see it on the map. Use the **Zoom To Selected Elements** tool on the **Draw** toolbar to see the text box.

3. Select the **Select Elements** tool and double-click on the text to open the **Properties** window. In the **Text** panel, type North Atlantic Ocean.

4. We would like to add a special character called Trade Mark Sign (Unicode character U+2122) next to the map text. To add a special character, we should access **Character Map**. A regular Windows operating system installation will add **Character Map** to your computer by default. Open your Windows **Control Panel**, go to **Appearance and Personalization | Fonts** (or <drive>:\ Windows\Fonts), and click on **Find a character** to open the **Character Map** window. Select **Advanced view**. For **Search for**, type trade and click on **Search**. Double-click on the character to add it to the **Characters to copy** text box. Click on **Copy**.

5. Go back to the ArcMap application. In the **Text** panel, click next to North Atlantic Ocean, and then right-click to select **Paste**. The TM symbol is shown as follow: North Atlantic Ocean TM.

6. Click on the **Change Symbol** button, and for **Color**, choose **Cretan Blue** (RGB: 0, 112, 255). Change **Font** to **Arial Rounded MT Bold**. For **Size**, type 200, and for **Style**, click on the Italic icon.

7. Click on the **Edit Symbol** button and select **Mask** tab. Select **Halo** and change the size to 20. Click on the **Symbol** button and for **Fill Color**, choose **Sodalite Blue** (RGB: 190, 232, 255) from the color palette window. Click on **OK** twice to return to the **Symbol Selector** window.

8. Save your text as My Ocean with **Category**: Water. Click on **OK** twice to close all dialog windows.

 From **Customize** menu, select **Style Manager** to inspect your default personal style file. The text symbol called My Ocean is stored in the Text Symbols folder.

9. On the **Tools** toolbar, click on **Go Back To Previous Extent**. Adjust the text position using the **Select Elements** tool on the **Draw** toolbar.

10. Next, we will add a graphic text using the **Label** tool on the **Draw** toolbar. The **Label** tool allows us to label a country from the Countries layer, based on its attribute value. To see and change the default attribute field used by the **Label** tool, open the **Layer Properties** window, go to the **Labels** tab, and inspect the field name from the **Label Field** drop-down list.

11. Select the **Label** tool, and click on the country, **Greenland**. In the **Label Tool Options** window under **Label Style**, select **Choose a style** and select the Country symbol. Close the dialog window. The text box is too small. Use the **Zoom To Selected Elements** tool to see the text box.

12. Select the **Select Elements** tool and double-click on the text to open the **Properties** window. Click on the **Change Symbol** button, and change **Size** to 200. Close all dialog windows. On the **Tools** toolbar, click on **Go Back To the Previous Extent**. Adjust the text position.

13. These two graphic texts are associated with the World data frame and stored as map document annotations in our Label.mxd map document. Right-click on the World data frame and select **Properties**. Click on the **Annotation Groups** tab, and note the default annotation group called <Default>. Deselect the default annotation group and click on **Apply** to update changes on the map.

14. Drag the **Data Frame Properties** window away so that you can see the map display. Note that graphic texts have disappeared (turn off). In the **Data Frame Properties** window, check the default annotation group again and click on **Apply** to turn on your graphic texts.

15. Save your map as MyLabels.mxd at ..\Chapter4\Labels. Leave the map document open to continue working with dynamic labels in the next section.

# Working with dynamic labels

In this section, we will use dynamic labels to automatically label all features in a layer at once. Then, we will convert the dynamic labels in a layer to a map document annotation to make them individually selectable in order to change their text characteristics.

The dynamic labels are displayed based on attribute values from the layer's attribute table or a related nonspatial table.

The ArcMap default label engine will determine the best placement of the labels on the available space on the map, and it will try to avoid label overlapping or features and labels overlapping. Therefore, the position and number of labels are dynamic, and they change as you change the map scale or data frame's extent (for example, using the **Pan** tool). For example, ArcMap will display all or almost all labels on the map at a large scale and fewer labels at a small scale.

The ArcMap application also allows you to label or set label text characteristics to particular groups of features in a layer.

The main label text characteristics are: font (serif, sans serif, and display), style (regular, bold, italic, and bold italic), size, color, character spacing (distance between letters), and line spacing (leading or the spacing between text lines).

If you want to customize the position and text type characteristics for a specific label in a layout, you should convert feature labels to annotation. The annotation has different behavior than dynamic labels in ArcMap. The annotation will always stay at the same position on the map and has a reference scale for its text size. You can change the position and the text characteristics of the annotation on a map individually.

Follow these steps to start working with dynamic labels and map document annotations:

1. Open your map document, `MyLabels.mxd`. In **Table Of Contents**, the checkbox to the left-hand side of the `PopulatedPlaces_Population` layer is dimmed. This tells you that the layer is not displayed on the map.

2. From the **Bookmarks** menu, select **Europe** to make this layer visible. Another option is to right-click on layer and select **Zoom To Make Visible**.

3. Change the scale to **1:10** million. For **Scale**, select or type `10,000,000` and press *Enter*. Use the **Pan** tool on the **Tools** toolbar to center the European continent in your map display.

4. Let's display the dynamic label for the `PopulatedPlaces_Population` layer. In **Table Of Contents**, right-click on the layer and click on **Label Feature**. The cities are labeled on the map. Inspect the labels at different scales using the **Zoom in, Zoom out** and **Pan** tools. Note that the label placement changes, and labels appear, disappear, or overlap with the city point symbols.

In the next steps, we will improve the text symbol and the placement for the labels.

5. In **Table Of Contents**, double-click on layer to open the **Layer Properties** window. Click on the **Label** tab. Under **Text Symbol**, change the color to **Tarragon Green** (RGB: `112, 168, 0`) and **Size** to `6`.

 Use the **Eye Dropper** tool to identify the RGB values for the city's green color on the map (navigate to **Customize | Customize Mode | Commands**).

6. Click on the **Placement Properties** button and select **Conflict Detection**. For the **Label** weight drop-down list, choose **Medium**. For the **Feature** weight drop-down list, choose **High**. Leave the **Place overlapping labels** option deselected. Click on **OK**.

7. In the **Layer Properties** window, select **Labels** tab. Set the **Scale Range** as follows: **Out beyond** to `1:10,000,000`, and **In beyond** to `<none>`. Click on **OK**. Inspect the labels. Note that label placement still changes and the label text does not overlap the city point symbols because we increase the point symbol weight to `High` and reduce the label weight to `Medium`.

We modified label properties for the entire `PopulatedPlaces_Population` layer. Next, we will set a different text size for all cities with population higher than 5,000,000 (the last two range classes). To have different label styles in the same layer, we should group features in two label classes so that we can assign them with different labeling properties, as shown in the following screenshot:

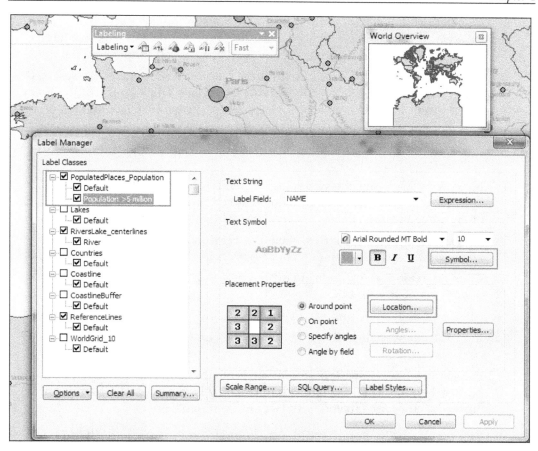

To group the features into label classes and assign them with different label properties, follow these steps:

8. From the **Customize** menu, navigate to **Toolbars | Labeling**. On the **Labeling** toolbar, click on the **Label Manager** tool. In the **Label Classes** panel, click on the **PopulatedPlaces_population** layer.

9. Under **Add label class**, for the **Enter** class name, type `Population >5 million` and click on the **Add** button. A new label class will be added to the **Label Classes** panel. The `Population >5` million label class inherited all the labeling properties from the `Default` label class.

10. The **Default** label class groups all the features in the layer and the
    `Population` >5 million label class as the same. This means that we have two
    labels for a feature on the map. Click on **Apply** to see changes on the map.

11. We will create an SQL query to limit the features in both label classes. In the
    **Label Classes** panel, select the **Default** label class. Click on the **SQL Query**
    button and build the `POP_MAX <= 5000000` expression. Click on **OK**.

12. In the **Label Classes** panel, select the **Population >5** million label
    class. Click on the **SQL Query** button and build the `POP_MAX > 5000000`
    expression. Click on **OK**.

13. Now, we will change the symbology of the labels. Under **Text Symbol**, change
    **Font** to **Arial Rounded MT Bold** and **Style** to **Bold**. Change **Size** to `10`. Click
    on **Symbol** and save the text symbol as `My City` with **Category**: `World`.

14. Under **Placement Properties**, click on **Location** and inspect the options.
    Leave the selected **Prefer Top Right** option unchanged.

15. Click on **Scale Range**, and set **Out beyond** to `20` million and **In beyond** to
    `3,000,001`. Click on **OK**.

16. Click on **Label Style** and save your label as `My City Label`. Click on **OK**.

17. In the **Label Manager** dialog window, click on **OK** to apply changes and
    close the window.

    Inspect the labels at scale 1:15 million. Note that only cities with a population
    higher than 5,000,000 are labeled. Inspect the labels at scale 1:10 million.
    Now, the cities are labeled with two different label styles.

18. You may save the **PopulatedPlaces_Population** layer as an individual layer
    file (`.lyr`). Remember that a layer file stores the symbology of features,
    including labels, and the path to the source data.

    In the next step, we will label the `RiversLake_centerlines` layer.

19. On the **Label** toolbar, click on the **Label Manager** tool. In the **Label Class**
    panel, select the **RiversLake_centerlines** layer. Click on **Apply** to see the
    updates on the map. Click on **OK** to close **Label Manager**. Inspect the river
    labels at scale 1: 5 million. From the **Bookmarks** menu, select **River**. Try to
    visualize and memorize some city and river label placements, as shown in
    the following screenshot:

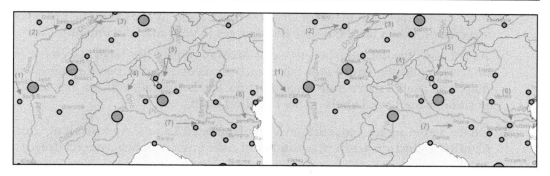

20. Open the **Label Manager** window again. Select the **RiversLake_centerlines** layer.

21. Based on the two categories of rivers from the symbol legend, **Label Manager** suggests that you add two corresponding label classes. We will label only river features. In the **Select symbology categories** list, uncheck **Lake Centerlines** and keep **River** checked. Click on the **Add** button.

22. In the pop-up window **Overwrite label classes?** click on **Yes**.

23. In the **Label Class** panel, select the **River** label class. For **Label Field**, retain the field name.

24. The text symbol has inherited the text color from the line symbol color called **My Water**. Change **Size** to 5. Click on the **Properties** button. Under **Orientation**, select **Curved**. From the **Orientation System** drop-down list, select **Line**.

25. Click on the **Conflict Detection** tab. For the **Label weight** drop-down list, choose **Low**. For the **Feature weight** drop-down list, choose **High**. Leave the **Place overlapping labels** option deselected. Click on **OK**.

26. Click on **SQL Query**, and note the default expression. ArcMap automatically created it for us based on the previous point symbol settings.

27. Click on **Scale Range**, and set **Out beyond** to 20 million and **In beyond** to 3 million. Click on **OK** twice to close all dialog windows.

28. On the **Labeling** toolbar, click on the **Label Priority Ranking** tool. Move the `PopulatedPlaces_Population-Population>5` million label class to the top of the list using the top up arrow button. Move the `PopulatedPlaces_Population-Default` and `RiverLake_centerlines-River` label classes to the second and third positions in the list. Click on **OK**.

The label classes with a higher priority will be placed first on the current map extent to the detriment of those with a lower priority.

29. Inspect the label placement for city and river labels, as shown in the previous screenshot. When finished, save the changes to your map document.

    As you saw, the default label engine of ArcMap controls the label placement and automatically redraws the labels each time you zoom or pan through your map. You cannot move individual dynamic labels. Instead, you can lock their position using **Lock Labels** on the **Labeling** toolbar.

30. On the **Labeling** toolbar, click on **View Unplaced Labels**. As a result of your **Conflict Detection** settings, a number of river and city labels weren't drawn.

In the next steps, we will add a new layer on the map and will label it based on two fields from a related table, as shown in the following screenshot:

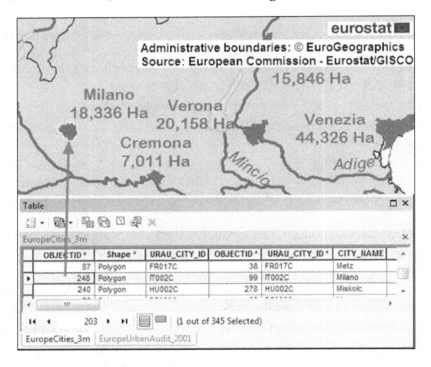

31. In **Table Of Contents**, turn off the `PopulatedPlaces_Population` layer. On the **Standard** toolbar, click on the **Add Data** button, navigate to `<drive>:\LearningArcGIS\Chapter4\World.gdb\Europe`, and select the **EuropeCities_3m** feature class. Click on **Add**.

32. Right-click on the layer and select **Open Attribute Table**. Inspect the attributes. The table doesn't contain the name of cities. We will use the user-added field, `URAU_CITY_ID`, to append additional data stored in a nonspatial table through the **Join** method.

33. Use the **Add Data** button again, navigate to `..\Chapter4\World.gdb`, and select the **EuropeUrbanAudit_2001** nonspatial table. Click on **Add**.

 Note that **Table Of Contents** lists the layers by source to see the table. In the **List By Drawing Order** mode, you cannot see a nonspatial table.

34. Right-click on table and select **Open**. In the **Table** window, the **EuropeUrbanAudit_2001** tab should be selected. Inspect the attributes. The table contains several useful fields, such as URAU_CITY_ID, CNTR_CODE, and CITY_NAME. To display the name of the cities and country code on the map, we need to use the **Join** method. The URAU_CITY_ID field is common to both tables **EuropeUrbanAudit_2001** and **EuropeCities_3m**. Leave the **Table** window open.

35. In **Table Of Contents**, right-click on the **EuropeCities_3m** layer and navigate to **Joins and Relates | Join**. The **Join attributes from a table** should be selected. From the **Choose the field in this layer that the join will be based on** drop-down list, select **URAU_CITY_ID**. From the **Choose the table to join to this layer** drop-down list, select **EuropeUrbanAudit_2001**.

36. From the **Choose the field in the table to base the join on** drop-down list, **URAU_CITY_ID** should be already selected. Accept the default **Join Options**. Click on the **Validate Join** button. Read the validation report and click on **Close** and then **OK**.

37. In the **Table** window, click on the **EuropeCities_3m** tab to inspect the attributes.

38. In **Table Of Contents**, right-click on the **EuropeCities_3m** layer, and select **Properties**. Click on the **General** tab, and set the scale range for layer as follows: **Out beyond** to 3 million and **In beyond** to <none>.

39. Click on the **Labels** tab. Check the **Label** features in this layer at the top-left corner. Click on the **Expression** tab, and build the `[EuropeUrbanAudit_2001.CITY_NAME] & vbnewline & FormatNumber([EuropeCities_3m.Shape_Area]/10000,0) & " Ha"` expression.

 We already built the expression for you. Click on the **Load** button, navigate to `<drive>:\LearningArcGIS\Chapter4\Labels`, and select `EuropeCities_3m_Label.lxp`.

40. Click on **Verify** to preview the label text. We used the CITY_NAME field from a nonspatial table and the Shape_Area field from the EuropeCities_3m layer. The FormatNumber function helps us convert meters to hectares.

41. Click on **OK** to return to the **Labels** in the **Layer Properties** window. Click on the **Symbol** button and select the **My City** symbol from your personal style. Click on **OK**. Change **Size** to 3, and deselect the **Bold** style.

42. For **Placement Properties**, select **Remove duplicate labels**. For the **Conflict Detection** tab, change the **Label weight** and **Feature weight** to **Medium**.

43. Set **Scale Range** for labels as follows: **Out beyond** to 3 million, and **In beyond** to <none>. Click on **OK** to close all dialog windows.

    Inspect the results of your previous settings on scale range for layer and labels using the following display scales: 1:4 million; 1: 3,000,001; 1:3 million, 1:2 million.

In the last part of this exercise, we will convert dynamic labels to map document annotations:

1. In **Table Of Contents**, right-click on the **World** data frame, and select **Convert Labels To Annotation**.

2. Under **Store Annotation**, select the **In the map** option. Notice that **Reference scale** for annotation will be 1:10 million, which is the data's reference scale.

3. Accept the default names for the two map annotation groups. Deselect the **Convert unplaced labels to unplaced annotation** option. We will have all unplaced labels in order to individually adjust their locations. Click on **Convert**.

4. First, right-click on the **PopulatedPlaces_Population** layer to see that **Label Features** is not selected. The labels on the map are now graphic texts stored as map document annotations.

5. Second, right-click on the **World** data frame and navigate to **Properties | Annotation Groups**. Inspect the two new annotation groups. Close the dialog window.

6. Inspect the results on the map. Note that annotations stay at the position where the dynamic labels were placed by ArcMap before you start converting them to map annotations. If you want to move or change every annotation manually, use the **Select Element** tool on the **Draw** toolbar or on the **Tools** toolbar.

7. If you work with dynamic labels in your map display again, deselect the corresponding annotation groups and check the **Label Features** from the layer's context menu again.

8. Save your map document as MyLabels.mxd at ..\Chapter4\Labels.

You can find the results of this exercise at <drive>:\LearningArcGIS\Chapter4\ Results\Labels\Labels.mxd.

# Creating a thematic map

In this section, we will map the density and proportion of the world's population.

You can map density using population density values or visually using dots to represent the number of people.

The first method implies the calculation of density values by dividing the quantity population values by the area of each country (data normalization) and symbolizing them with graduated colors or symbols.

The second method is called a **dot density map**. Each dot represents a given number of people and its position is not associated with the real location. ArcMap randomly places the corresponding number of dots within each polygon feature, representing a country.

You can map the proportion of the world's population by dividing (normalizing) the population of each country by total population. The proportion can be represented in the legend as ratios (values between 0 and 1), percentages (ratio multiplied by 100), or rates (per person, per 1,000 persons), and they can be symbolized with graduated colors or symbols.

Follow these steps to start creating two quantitative thematic maps:

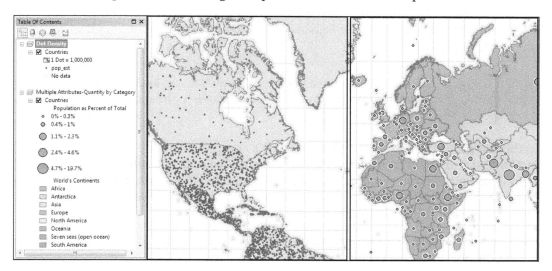

In the first part of this exercise, we will symbolize the Countries layer using dot density, based on the pop_est field:

1. Start the ArcMap application to open an existing map document, ThematicMaps.mxd, from: ..\Chapter4\ThematicMap.In **Table Of Contents**, right-click on the **Country** layer and navigate to **Properties | Symbology**.

2. In the **Show** area, click on **Quantities**, and select **Dot Density**. From the **Field Selection** list, select **pop_est** and click on the right arrow button.

3. Double-click on the point symbol for pop_est and change the color to a darker red. Click on **OK**. In the **Background** section (lower right of the window), click on the line symbol and change the width to 0.1. Click on **OK**. Click on the color box to change the background color to **Mango** (RGB: 255, 211, 127).

4. Click on the **Exclusion** button, and build the pop_est = -99 expression. Click on the **Legend** tab and select **Show symbol for excluded data**. Change the color to yellow. For **Legend**, type No data. Return to the **Layer Properties** window. Accept the default option for **Maintain Density by Dot Value**. When the map scale changes, the dot size will not be scaled (increased or decreased).

5. Let's adjust the value of the dots. Make sure that **Dot Size** is 2. In **Dot Value**, type 1,000,000 and press the *Tab* key. The **Min**, **Mean**, and **Max** boxes have updated. Click on **OK** to apply changes and close the window.

6. Turn on all layers beneath the **Countries** layer. Inspect the results on the map at different scales. When finished, from the **Bookmarks** menu, select **Density of Population**.

> Use the *F5* key to redraw the layers on the map or click on the small **Refresh (F5)** button at the bottom-left corner of the map display area.

In the last part of this exercise, we will create a complex thematic map by combining qualitative and quantitative data. As background, we will symbolize the countries with their own unique symbol, based on the qualitative field called continent. The foreground will display the proportion of population relative to the world's total population using graduated circular symbols:

7. First, let's add a new data frame. From the **Insert** menu, select **Data Frame**. Your data frame is blank because it has no layer.

> Note that the default name is bolded. This tells you that the data frame is the active data frame. In a map document, only one data frame can be active at a time.

8. Change the data frame's name to `Multiple Attributes-Quantity by Category`.

9. Secondly, we will add a layer. Right-click on the **Country** layer from the **Dot Density** data frame and select **Copy**. Right-click on the active data frame and select **Past Layer(s)**.

10. Third, we will symbolize the layer. Double-click on the layer to open **Layer Properties**. The **Symbology** tab should be selected.

11. In the **Show** area, navigate to **Multiple Attributes | Quantity by Category**. For **Value Field**, select the qualitative field called `continent`. Deselect `<all other values>` and click on **Add All Values**. Change the legend's heading title to `World's Continents`.

12. From **Color Scheme**, select **Cool Tones**. Under **Variation by**, click on the **Symbol Size** button to symbolize the foreground based on the quantitative `population` field, normalized by the total population.

> Do not use more than 12 different colors and more than eight distinct shades of the same color on a thematic map because of the human eye's limitation in deciphering large number of colors. Source: `http://video.esri.com/watch/77/basic-principles-of-cartographic-design`.

13. Under **Fields**, from the **Value** drop-down list, select **pop_est**. From the **Normalization** drop-down list, select **<PERCENT OF TOTAL>**.

We will accept the `Natural Breaks (Jenks)` classification scheme. Inspect the classification histogram. The histogram has one prominent peak and the proportion of population values are strongly skewed to the right. Accept all five default classes. If there are too many classes, the symbols will became indistinguishable and the circles symbols of the high values will obscure each other.

14. Click on the **Classify** button, select the **Exclusion** button, and build `pop_est = -99` expression. For **Legend**, set the color of the symbol to red and **Label** to `No data`.

15. Return to the **Draw quantities using symbol size to show relative values** window. Accept the default **Symbol Size from** `4` **to** `18`. Click on the **Template** button to change the point symbol to `Circle 1` and color to `Medium Coral Light`. Navigate to **Edit Symbol | Mask** and click on **Halo**. Accept the default color (white) and size.

16. Return to the **Draw quantities using symbol size to show relative values** window. Click on the **Label** column heading and select **Format Labels**. Set the **Number of decimal places** to 1. Return to the **Layer Properties** window.

17. You may label the countries or select the **Show MapTips using the display expression** option in the **Display** tab. Click on **OK**.

18. In **Table Of Contents**, change the legend title to Population as Percent of Total. Turn on all layers beneath the Countries layer for a nice-looking map.

19. Inspect the results on the map at different scales. When finished, from the **Bookmarks** menu, select **Proportion of Population relative to the World's Total Population**.

Finally, we will use **Layout View** to see the two thematic maps at the same time in a layout:

20. From the **View** menu, select **Layout View**. Note that a new toolbar will automatically be displayed. The two data frames are overlapping. To select, drag, and resize a data frame in **Layout View**, use the **Select Elements** tool on the **Tools** toolbar.

21. In **Layout View**, the selected data frame is outlined in blue and has eight selection handles. Look at **Table Of Contents** to see which data frame you have selected (the selected data frame has its name in bold).

22. If you want to zoom in **Layout View**, exclusively use the **Zoom** and **Pan** tools from the **Layout** toolbar. If your symbols are not correctly drawn, use the **Refresh** tool from **View**. You will learn more about how to create a map layout in *Chapter 10, Designing Maps*.

23. Save your map document as MyThematicMaps.mxd at ..\Chapter4\ ThematicMap.

You can find the results of this exercise at <drive>:\LearningArcGIS\Chapter4\ Results\ThematicMap.

# Summary

In this chapter, you combined different visual variables (size, shape, or hue) and feature types (point, line, and area) to represent the geographic object on the map. In addition to this, you explored different ways to label features in the data frame and map layout. You also learned how to add a special character to a graphic label text using the **Character Map** Windows collection.

In the second part of this chapter, you saw how the symbolization method, classification scheme, and number of classes are influenced in the message of your thematic map.

In the next chapter, you will learn how to create and edit feature shapes and attributes using the ArcMap editing tools.

# Creating and Editing Data

# 5

In this chapter, we will work with the ArcGIS for Desktop editing tools to create and edit feature shapes and attributes. We will describe the main steps in the feature editing process to use two different methods to edit feature attributes and create point features using tabular data.

By the end of this chapter, you will learn about the following topics:

- Editing feature shapes and attributes using ArcMap editing tools
- Maintaining spatial relationships among features using map topology
- Spatially adjusting vector data to real-world units
- Editing and calculating feature attributes
- Creating point geometry from $\varphi$, $\lambda$ coordinates

## Creating and editing features

In this chapter, we will work with three primary features' geometries: point, line, and polygon. A point feature is defined by a single pair of $x$, $y$ coordinates. Line and polygon features are defined by vertices and segments.

The vertices represent ordered $x$, $y$ coordinate pairs that form the feature's shape. The coordinates of a vertex can also include M (measure) and Z (elevation) values.

A segment connects only two vertices. A line with two or more segments is called a **polyline**. ArcGIS treats lines and polylines as basically the same `Line feature` feature type, and they can be stored in the same feature class.

In ArcGIS, the lines and polygon feature classes can store single or multipart features. A multipart feature is composed of two or more disconnected shapes that have only one record in the feature class attribute table. A simple point feature class cannot store multipart features, but a *multipoint feature class* can store multipoint features.

In the editing process, the shape of a feature can be modified by moving, adding, or deleting the vertices that form its sketch. A sketch displays the location of vertices and segments that define the feature's shape.

There are six main steps in the feature's shape-editing process:

1. Adding the data to a map document.
2. Symbolizing layers as desired.
3. Starting an edit session.
4. Setting a feature template and the snapping properties to edit the layer's feature geometry.
5. Creating a new feature by adding a new sketch or selecting an existing feature and displaying its sketch.
6. Editing the feature shape by moving, deleting, or inserting one or more vertices.
7. Editing the attribute values for edited or created features.
8. Saving the feature edits and stopping the edit session.

# Editing features

In this section, you will learn how to create a new feature class from an existing one using the **Export Data** tool in the ArcMap application. Next, you will explore the different edit tools used to create new features by splitting or combining existing features. In the end, you will learn how to modify the feature's shape by adding, moving, and deleting vertices on its sketch.

Before you start editing a feature sketch, you should always set the snapping properties, such as the snap type or the snapping tolerance value. Snapping helps you create coincidences during the editing process. Snapping allows you to move the mouse pointer exactly to a vertex, edge, start point, or endpoint of a feature when the pointer location is within a predefined distance of them. This distance is called **snapping tolerance**. ArcMap calculates the value of the snapping tolerance for you, but you can change this and specify your own value.

Follow these steps to start editing features using the ArcMap application:

1. Start the ArcMap application and open a new map document. Open the **Catalog** window, and navigate to `..\Chapter5\EditingFeatures\World.gdb`.

2. Expand the `Europe` feature dataset. Select the **EuropeCities_10k** feature class and drag it on the ArcMap map display. The map shows the cities of Paris and Brussels at scale 1:10,000.

3. In **Table Of Contents**, right-click on the **EuropeCities_10k** layer and select **Open Attribute Table**. Dock the **Table** window to the bottom of the map display by dragging and dropping it on the down arrow of the blue-centered target.

4. Click on the **Table Options** button at the top-left corner and select **Select by Attributes**. Build the `CITIES = 'Bruxelles/Brussel'` expression. Click on **Apply** and then click on **Close**. Note that 32,945 records are selected in the attribute table and on the map display. We will save these selected features to a new feature class called `Brussels`.

5. Right-click on the `EuropeCities_10k` layer and select **Export Data** from **Data**. For **Export**, accept the **Selected features** option. Click on the browser button, make that sure you look in the `..\Chapter5\EditingFeatures\World.gdb\Europe` folder, and type `Brussels` for the name of the feature class. Click on the **No** button to not add the exported data to map display as a layer. In the **Catalog** window, the `Europe` feature dataset has been updated with your new polygon feature class. Hide the **Catalog** window by clicking on the **Auto hide** button. Close the **Table** window.

Next, we will change the projection of the `Brussels` feature class and save it to the `Brussels` feature dataset:

6. On the **Standard** toolbar, click on the **ArcToolbox** tool. Navigate to **ArcToolbox | Data Management Tools | Projections and Transformations** and select **Project**. Set the following parameters:

   ° **Input Dataset or Feature Class**: Set this to `..\EditingFeatures\World.gdb\Europe\Brussels`

   ° **Output Dataset or Feature Class**: Set this to `..\EditingFeatures\World.gdb\Brussels\BrusselsCity`

   ° **Output Coordinate System**: Set this to `Belge_Lambert_1972`

   ° **Geographic Transformation**: Set this to `Belge _1972_To_ ETRS_1989_2`

7. Click on **OK**. Click on **No** to not add the exported data to map display as a layer. Hide the **ArcToolbox** window by clicking on the **Auto hide** button.

8. On the **Standard** toolbar, click on the **New** tool to open a new map document. In the **New Document** dialog window, for **Default geodatabase for this map**, click on the browser button and navigate to <drive>:\LearningArcGIS\ Chapter5\EditingFeatures\World.gdb. The World.gdb file geodatabase will be the default location if you use different editing and geoprocessing tools which create new output feature classes in ArcMap. Click on **Add** and then click on **OK**. Do not save changes from the last map document.

9. Add the **BrusselsCity** and **Buildings** feature classes from the **Brussels** feature dataset in ArcMap by dragging them to the map display.

10. Change the **BrusselsCity** symbol to Hollow, **Outline Width** to 1.5, and **Outline Color** to Medium Coral Light. Label these features with the OBJECTID field's values. Change the label color to Medium Coral Light.

11. Change the **Buildings** symbol to Hollow, **Outline Width** to 1.5, and **Outline Color** to Solar Yellow. Label the **Buildings** features with the OBJECTID field's values.

12. Use **Add Data From ArcGIS Online** in **Add Data** to add two aerial photos of Brussels: Brussels -Othotophoto 2004 and Brussels -Othotophoto 2012. In **Table Of Contents**, drag Brussels -Othotophoto 2004 to the bottom of the list.

13. From the **Bookmarks** menu, select **Manage Bookmarks**. Click on the **Load** button, navigate to <drive>:\LearningArcGIS\Chapter5, and select the EditingFeaturesAttributes.dat ArcGIS Place file. Click on **Open**. These bookmarks will help us identify particular study areas in the next exercises. Click on **Close**.

14. When finished, save your map document as MyEditingFeatures.mxd to <drive>:\LearningArcGIS\Chapter5\EditingFeatures.

In the next steps, we will start editing the building footprints from the Buildings layer. These buildings were digitized using an aerial photo from 2004 (Brussels -Othotophoto 2004) as the layer's reference. We will update these features to reflect the changes that were made from 2004 to 2012 using the Brussels -Orhotophoto 2012 image as a reference layer.

15. First, we will set the Buildings layer as the only selectable layer in the map. At the top of **Table Of Contents**, click on the **List By Selection** button. Both the BrusselsCity and Buildings layers are selectable. Click on the **Selectable** button next to the **BrusselsCity** layer to move it to the **Not Selectable** list. Return to the **List By Drawing Order** layers list mode.

16. When editing layers, it is good practice to keep your **Table Of Contents** set to the **List By Drawing Order** list mode. This allows you to see what layers are selectable, control the selectable layers, see which layers have features selected, and how many feature are selected. This can prevent you from accidentally deleting or changing features that you do not wish too during an editing session. From **Bookmarks**, select **Edit 1**. You will divide a feature into five separate features, as shown in the following screenshot:

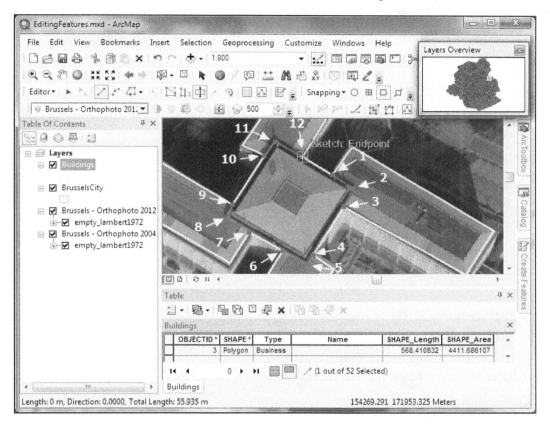

17. In **Table Of Contents**, right-click on the **Buildings** layer and select **Open Attribute Table**. Dock the **Table** window to the bottom of the map display.

18. First, we will start an edit session and set the snapping properties. If necessary, add the **Editor** toolbar. On the **Editor** toolbar, select **Start Editing** in **Editor**.

19. We will use the default snapping environment to enable snapping to feature vertices. From the **Editor** menu, select **Snapping Toolbar** in **Snapping**. On the **Snapping** toolbar, turn off (deselect) all snapping types except for the **Vertex Snapping** button that should be selected. During the edit process, your mouse pointer will be moved exactly to the vertices of a feature.

> If you need more control of snapping properties, such as the individual type of snapping on each editable layer or snapping tolerance units (pixels or map units), ArcMap allows you to activate the classic snapping environment that was used by older versions of ArcGIS. To activate the classic snapping, on the **Editor** menu, navigate to **Editor | Options | General** and select **Use classic snapping**. From the **Editor** menu, select **Snapping Window** in **Snapping** to manage the individual snapping on layers. To change the snapping tolerance units, navigate to **Editor | Snapping** and select **Options**.

Next, we will create a *feature template* that defines all settings that are necessary to create a new feature, such as layers, symbology, and default attribute values, and the default construction tools used to create it.

20. If the **Create Features** window isn't open, you should navigate to **Editor | Editing Windows** and select **Create Features**. As the BrusselsCity and Buildings layers reference two feature classes that are stored in the same workspace called World.gdb, the **Create Features** window displays the default feature templates for both of them.

> During the edit session, you can modify the properties of an existing feature template or create a new feature template.
>
> A feature template is saved as a property of the map document (.mxd) and layer file (.lyr).

21. In the **Create Features** window, double-click on the default Buildings template to open the **Template Properties** window. Inspect the information for the template displays. For **Description**, type Brussels buildings. For **Default Tool**, accept Polygon. Click on **OK** to update the Buildings feature template. At the bottom of the **Created Features** window, in the **Construction Tools** list, make sure that the Polygon tool is selected.

22. On the **Editor** toolbar, click on **Edit Tool**. At the bottom of the **Table** window, click on the **Show selected records** button to view only the selected building record.

23. On the **Editor** toolbar, select the **Cut Polygons** tool. Note that another tool called **Straight Segment** has been selected. You will cut the polygon feature using a multipart polyline.

24. Move the pointer close to the start point, **1**, indicated in the previous screenshot, to view the SnapTip called Buildings: Vertex. Click on the point, **1**, to add the first vertex of the segment. Click on the point, **2**, and again on the point, **3**. Right-click on the third vertex and select **Finish Part**.

   If you want to delete a vertex, right-click on it and select **Delete Vertex**.

   You can use the **Zoom in, Zoom out**, and **Pan** tools while you are splitting the polygon, but you have to select the **Cut Polygons** tool again to continue adding vertices.

25. Click on the points **4, 5**, and **6** as shown in the previous screenshot. Right-click on the sixth vertex and select **Finish Part**. Repeat these steps for the rest of the points. Right-click on the twelfth vertex (Sketch: Enpoint) and select **Finish Sketch**.

26. Five new features have been created and selected on the map display. In the **Table** window, notice that the Buildings feature class table has been updated with five new building records that correspond to the five polygons in the map display. To deselect these new features on the map, use the **Clear Selected Features** tool on the **Tools** toolbar.

27. To save your edit to the data, select **Save Edits** from the **Editor** menu.

In the next steps, we continue to update the buildings by removing two buildings that have been demolished since 2010, and we will merge two existing features to represent a larger commercial building, as shown in the following screenshot:

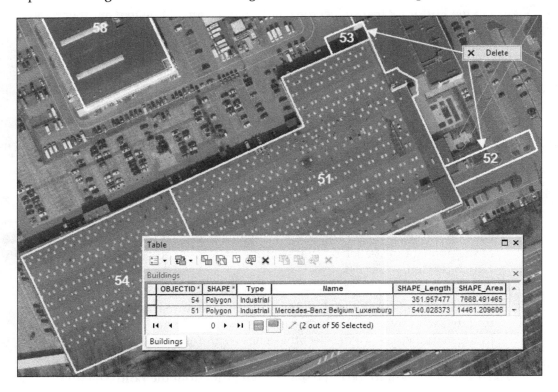

28. We will continue to use the Buildings feature template. From **Bookmarks**, select **Edit 2**.

29. Add the **Effects** toolbar to the map display using the same method that you used to add the **Editor** toolbar, and select **Brussels-Orthophoto 2012** from the drop-down list of layers. Select the **Swipe** button. Click on the map display and drag to see the differences between the two orthophoto maps.

30. Use the **Select Feature** tool and the *Shift* key to select two buildings that have the following OBJECTID values: 53 and 52. Right-click on the map display and select **Delete**. You have a second option to remove them from the Buildings layer using the **Delete Selected** tool in the **Table** window.

During an edit session, you should use the **Edit** tool to select a feature by clicking on it. As we don't set **Sticky move tolerance** by navigating to **Editor | Options | General**, we used the **Select Feature** tool to prevent accidental moving of the selected features.

For more information about sticky tolerance, please refer to ArcGIS Resource Center at http://resources.arcgis.com/en/help/. Navigate to **Desktop (ArcMap): 10.4 | Manage Data | Editing | Editing existing features | Preventing inadvertent movement of features when selecting**

31. Use the **Select Feature** tool to select the buildings with the OBJECTID values, 51 and 54. In the **Table** window, inspect the attribute values of the building with the ID, 51.

32. On the **Editor** toolbar, navigate to **Editor | Merge**. In the **Merge** dialog window, click on the first feature, Industrial (Buildings). The selected feature flashes on the map. As the first feature in the list is the larger one and has a name, we would like to keep its attribute values. Click on **OK**.

33. In the **Table** window, notice that the OBJECTID, Type, and Name values were inherited from the previous selected feature. Instead, the SHAPE_Length and SHAPE_Area fields have been updated by ArcMap.

In the next steps, we will reshape the building with OBJECTID 51, as shown in the following screenshot:

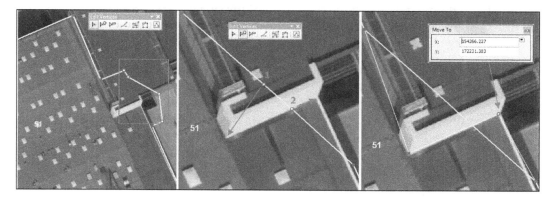

34. To edit the building shape, you must display its sketch. On the **Editor** toolbar, select the **Edit Tool** button, and double-click on the building to select it and display the feature's sketch, as shown in the previous screenshot. The **Edit Vertices** mini-toolbar will automatically be added on the map.

35. Drag and draw a box over the four vertices, as shown in the previous screenshot. Move the pointer over one of the selected vertices until it turns into a compass arrow, then right-click on the selected vertex and select **Delete Vertices**. If you make a mistake, use the **Undo Delete** tool from the **Standard** toolbar and repeat the step. To erase two or more sketch's vertices, you can also draw a box over the vertices using the **Delete Vertex** tool from the **Edit Vertices** mini-toolbar.

36. Click away from the selected feature to deselect it and see the changes. On the **Editor** toolbar, select the **Edit Tool** button again and double-click on the building to display the feature's sketch. On the **Edit Vertices** toolbar, select **Add Vertex**. Click on the line segment and add two vertices, as shown in the previous screenshot.

37. Click on the first new vertex (**1**) to select it. Move the pointer over the selected vertex until it turns into a compass arrow. Drag and drop the vertex to the new location.

38. Click on the second new vertex (**2**) to select it. You will move this feature's vertex to an exact *x*, *y* location. Right-click on vertex and select **Move To**. First, click on the down arrow next to the **X** box, and check whether **Meter** is selected. In the **X** box, select and delete the existing value. Type 154266.227 and press the *Tab* key. For **Y**, type 172231.383 and press the *Enter* key. Note that vertex has moved to the exact location. Press the *F2* key to finish the sketch.

    Even if your feature shape has changed in your map document, the edits of the data are not saved in your feature class. Saving your map document does not save the data edits to your geodatabase.

39. To store the updated date to your file geodatabase, you must save your edits during an edit session or when you stop the edit session. On the **Editor Toolbar**, navigate to **Editor | Save Edits**. ArcMap does not include an autosave function, so you need to save your changes often. Also, there is no way to recover unsaved edits if something happens before you save.

In the last steps, we will move three adjacent buildings and will reshape one of them, as shown in the following screenshot:

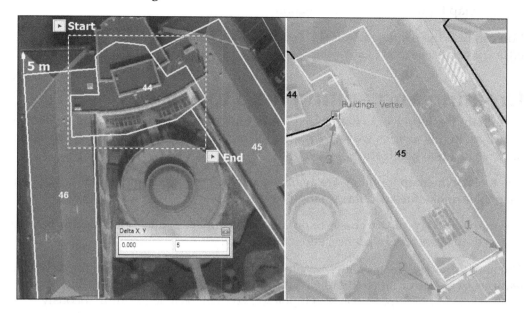

40. From **Bookmarks**, select **Edit 4**. On the **Editor** toolbar, click on the **Edit** tool. Drag and draw a box over the three buildings, as shown in the previous screenshot. Note that the `Start` and `End` points of the box are not touching the polygon features. This way you can select three adjacent buildings with the **Edit** tool without accidentally making small position moves.

41. To move all buildings 5 meters on the **Y** axis, from the **Editor** toolbar, navigate to **Editor | Move**. In the **Delta X, Y** window, type 5 in the second box and press the *Enter* key. All three buildings have been shifted by 5 meters on the **Y** axis.

42. Next, we will adjust the shape of a building. On the **Tool** toolbar, click on the **Clear Selected Features** tool to deselect the buildings. With the **Edit** tool, select the building with `ID 45` and then click on the **Reshape Feature** tool on the **Editor** toolbar.

43. Click on the point, **1**, to add first vertex of the line segment. Click on the point, **2**, to add the second vertex, and then click on the point, **3**, to add the third vertex. Press the *F2* key to finish the sketch. Inspect the results. Deselect the building and save the edits. On the **Editor** toolbar, stop the edit session by navigating to **Editing | Stop Editing**.

44. When finished, save your changes to the map document. Leave the `MyEditingFeatures` map document open to continue working with map topology in the next section.

You can find the results at `<drive>:\LearningArcGIS\Chapter5\ Results\ EditingFeatures\EditingFeatures.mxd`.

# Creating and editing map topology

ArcGIS for Desktop works with two types of topology: map topology and geodatabase topology. In this section, we will use only the map topology tools to edit the coincident features in a polygon feature class.

The tools of the map topology are available at all license levels and will also work with shapefiles unlike the geodatabase topology.

The map topology allows you to maintain three primary spatial relationships between features: adjacency, coincidence, and connectivity.

The map topology creates temporary topological relationships between the coincident features from one or more editable layers in the map document. By selecting two or more layers to participate in a map topology, you will be able to simultaneously edit the shape of all coincident features from these layers. The topological relationships are maintained only during an edit session and are identified on-the-fly in the current map extent.

In order to assure the spatial relationships between features, map topology works with the **cluster tolerance**. The cluster tolerance is synonymous with **XY Tolerance** term. You have worked with **XY Tolerance** in *Chapter 3, Creating a Geodatabase and Interpreting Metadata*. Esri documentation defines the cluster tolerance as "the minimum distance between coordinates before they are considered equal". In the map topology context, the Esri GIS Dictionary defined the cluster tolerance as "the minimum tolerated distance between vertices" (*A to Z GIS: An Illustrated Dictionary of Geographic Information Systems, Tasha Wade and Shelly Sommer, Esri Press*).

When you edit features in a map topology, you work with three topology elements: edge, node, and pseudo-node. An edge is the shared segment(s) between two adjacent features. A node connects at least three shared segments between two or more adjacent polygon features. The nodes are the beginning and ending points of an edge. A pseudo-node connects two segments that share the same attribute values or is the node where a single polyline connected with itself. With ArcMap topology editing tools, you can move nodes and move, reshape, or modify edges.

Follow these steps to start editing features using map topology in the ArcMap application:

1. Open your map document called `MyEditingFeatures.mxd` from `<drive>:\LearningArcGIS\Chapter5\EditingFeatures`.

2. From **Bookmarks**, select **Edit 5**. You will reshape two adjacent building features that share an edge, as shown in the following screenshot:

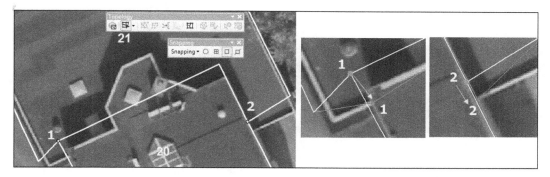

3. On the **Editor** toolbar, select **Start Editing** from the **Editor** menu. Click on the **Editor** menu again and navigate to **More Editing Tools | Topology**. Dock the **Topology** toolbar to the ArcMap window.

4. First, we should create a topology. On the **Topology** toolbar, click on the **Select Topology** tool. Select the **Buildings** layer. Click on the **Options** down arrow to inspect the value of the cluster tolerance. Accept the default value of `0.001` meter. Click on **OK** to create the topology.

5. Second, we should set the topology snapping properties. On the **Snapping** toolbar (**Editor | Snapping | Snapping Toolbar**), select the **Snapping** menu and click on **Snap To Topology Nodes**. Your mouse pointer will snap to a node in the map topology.

6. To see the topology nodes, from the **Editor** menu, navigate to **Options | Topology** and select **Unselect Nodes**. Click on **OK**. Note the nodes are displayed as black small circles.

7. On the **Topology** toolbar, click on the **Topology Edit** tool. Double-click on the first point, as shown in the previous screenshot. ArcMap builds the topology cache for the current map display and displays the selected nodes in magenta. If you see the building's sketch instead of the selected node, click away from the edge to deselect it and try again.

8. With the **Topology Edit** tool active, move your pointer over the selected node until it changes to a four-head arrow, click and drag it on the building's edge, as shown in the previous screenshot. Repeat this step for the second node.

9. With the **Topology Edit** tool active, click on the shared edge to select it, as shown in the following screenshot:

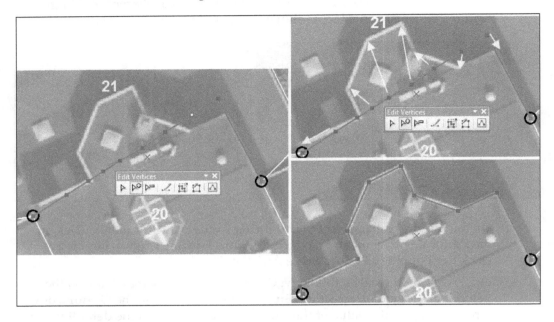

10. The selected edge is displayed in magenta, also. From the **Topology** toolbar, select the **Modify Edge** tool and click on the edge again to see the sketch's vertices.

11. We would like to add seven vertices on the shared edge to redraw it, as shown in the previous screenshot. From the **Edit Vertices** mini-toolbar, select the **Add Vertex** tool and start adding vertices by clicking on the selected edge.

12. With the **Topology Edit** tool active, move your pointer over the first new vertex until it changes to a four-head arrow, click and drag it on the building's corner, as shown in the previous screenshot.

13. Repeat these steps for the rest of vertices. Press the *F2* key to finish the sketch and deselect the edge.

14. On the **Editor** toolbar, save the edits by selecting **Save Edits** in **Editing**.

In the second part of this exercise, we will edit the adjacent polygons of the BrusselsCity layer using the map topology tools. Follow these steps to start editing the adjacent features:

15. First, we will set the BrusselsCity layer as the only selectable layer in the map. At the top of **Table Of Contents**, click on the **List By Selection** button. Both the BrusselsCity and Buildings layers are selectable. In the **Selectable** list, click on the **Selectable** button next to **Buildings** to move it to the **Not Selectable** list.

16. In the **Not Selectable** list, click on the **Selectable** button next to the **BrusselsCity** layer to move it to the **Selectable** list. Return to the **List By Drawing Order** layers list mode.

17. From **Bookmarks**, select **Edit 6**. You will reshape two shared edges, as shown in the following screenshot:

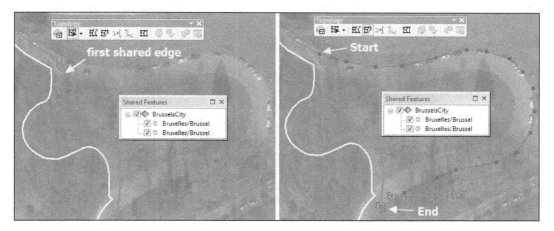

18. We should modify the map topology. On the **Topology** toolbar, click on the **Select Topology** tool. Deselect the **Buildings** layer and select the **BrusselsCity** layer. Click on **OK** to create the topology for the BrusselsCity layer.

19. On the **Topology** toolbar, click on the **Topology Edit** tool and select the first shared edge between a road and a green urban area. The topology cache is building, and the selected edge is displayed in magenta.

20. Right-click on the selected edge and select **Select Shared Features**. On the **Topology** toolbar, click on the **Shared Features** tool to see the list of the selected features.

If you want to see the values of the **ITEM** field in the **Shared Features** window, right-click on **BrusselsCity** and select **Display** from the **Properties** option. Under **Display Expression**, select the **ITEM** field, and click on **OK**. Close the **Shared Features** window and open it again.

21. On the **Topology** toolbar, click on the **Reshape Edge** tool. With the **Reshape Edge** tool activated, you can reshape the selected edge by starting to add vertices, as shown in the previous screenshot. After you add the last vertex, press the *F2* key to finish the sketch. Note that both the road and green urban areas were simultaneously updated.

22. Click on the **Topology Editor** tool again and select the second shared edge between the same road and an urban fabric area, as shown in the following screenshot:

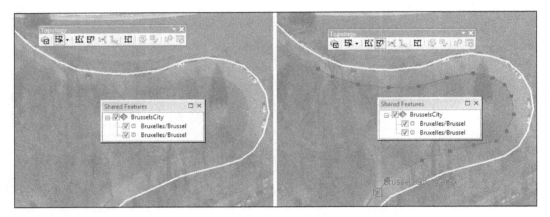

23. Activate the **Reshape Edge** tool, and redraw the shared edge, as shown in the previous screenshot. Press the *F2* key to finish the sketch. The shared edge has been updated. On the **Tools** toolbar, click on the **Clear Selected Features** tool to deselect the two adjacent features.

24. Stop editing and save your edits. When finished, save your changes in the MyEditingFeatures.mxd map document and close ArcMap.

You can find the results at <drive>:\LearningArcGIS\Chapter5\Results\ EditingFeatures\MapTopology.mxd.

# Using spatial adjustment

In this section, we will perform a spatial adjustment for two feature classes imported from a CAD file that doesn't have an associated real-world coordinate system.

To scale, rotate, and translate the data coming from the CAD file so that it closely aligns with your existing data, we will use the **affine transformation** method that requires at least three displacement links. A displacement link defines the source and destination coordinates that are used by the transformation method. Before starting any spatial adjustment, be sure that the horizontal accuracy of the destination coordinates is at least the same or better than the horizontal accuracy of the source coordinates.

To perform a spatial adjustment, we will use the **Spatial Adjust** toolbar that works on layers that only reference a shapefile or a geodatabase feature class. It will not work on the unconverted CAD data.

Follow these steps to perform a spatial adjustment in the ArcMap application:

1. Start the ArcMap application and open the existing `SpatialAdjustment.mxd` map document from `<drive>:\LearningArcGIS\Chapter5`.

 Please remember that the `Polyline` and `Polygon` layers were imported from a CAD file to the `World` geodatabase in *Chapter 3, Creating a Geodatabase and Interpreting Metadata*. As we don't have more metadata information from CAD Mapper, we assume that the reference scale for the data is 1:15,000.

2. In **Table Of Contents**, right-click on the **Polyline** layer and navigate to **Properties | Source**. Note that there is no projection information associated with the layer. Close the **Layer Properties** window. Repeat the step for the `Polygon` layer.

First, we will add a base map for the city of Brussels from ArcGIS Online. Check whether the ArcGIS for Desktop is connected to ArcGIS Online (see the *Using a coordinate reference system* section in *Chapter 2, Using Geographic Principles*).

3. On the **Standard** toolbar, from the drop-down next to the **Add Data** and click **Add Data from ArcGIS Online**. Search for the `brussels` data and select **Brussels - Orthophoto 2012**. Click on **Details** to read the information from the **Description** tab. Click on the yellow button called **Add Data**, as shown in the following screenshot:

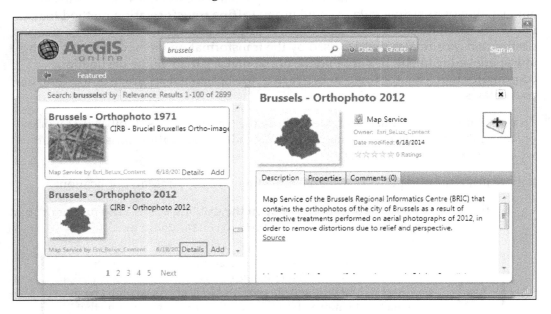

4. Drag the **Brussels - Orthophoto 2012** basemap layer to the bottom of **Table Of Contents**.

5. Right-click on the **Brussels-Orthophoto 2012** basemap layer, and navigate to **Properties | Source**. Inspect the resolution and coordinate system of the orthophoto map. The layer's coordinate system is `Belge_Lambert_1972` and the map projection is `Lambert_Conformal_Conic`. You have already worked with the `Belge_Lambert_1972` coordinate reference system in *Chapter 2, Using Geographic Principles*.

Second, we will use the **Spatial Adjustment** tools to move the `Polyline` and `Polygon` layers to their correct coordinates using the `Brussels-Orthophoto 2012` basemap layer as the destination coordinate space, as shown in the following screenshot:

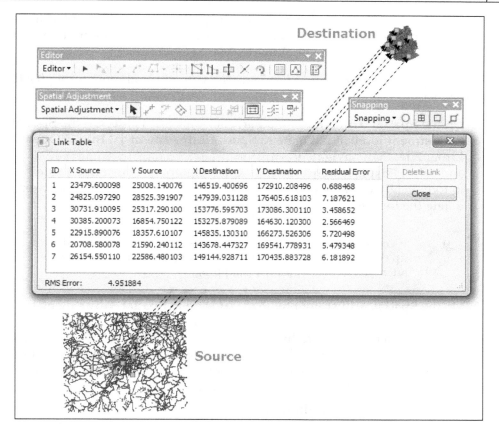

6. Add the **Spatial Adjustment** and **Editor** toolbars from **Toolbars** in **Customize**. To apply a spatial adjustment to the `Polyline` and `Polygon` layers, you must start the editing session. On **Editor Toolbar**, navigate to **Editor | Start Editing**.

7. On the **Spatial Adjustment** toolbar, navigate to **Spatial Adjustment | Set Adjust Data**. Select **All features in these layers** and make sure that both the `Polyline` and `Polygon` layers are selected. Click on **OK**.

8. On the **Spatial Adjustment** toolbar, navigate to **Spatial Adjustment | Adjustment Methods**. We will use the default transformation method called `Transformation-Affine`.

9. Go to the **Bookmarks** menu and select **Manage Bookmarks**. Click on the **Load** button and add `Source_points.dat` and `Destination_points.dat` from `<drive>:\LearningArcGIS\Chapter5`.

10. Keep **Bookmarks Manager** open, and inspect **Source** and **Destination** for all seven points by selecting the bookmark and clicking on **Zoom To button**. When finished, close the **Bookmarks Manager** window.

Instead adding the displacement links manually, you will load them through a links file, as follows:

11. On the **Spatial Adjustment** toolbar, click on **Link Table**. On the **Spatial Adjustment** toolbar, navigate to **Spatial Adjustment | Links | Open Links File**, then navigate to `<drive>:\LearningArcGIS\Chapter5`, and select **DisplacementLinks.txt**. Click on **Open**. Your **Link Table** automatically updates with seven coordinate pairs. You can modify or delete the individual links.

 If you want add the displacement links manually, you should set up the snapping environment by selecting the **Vertex Snapping** after navigating to **Editor | Snapping**.

12. On the **Spatial Adjustment** toolbar, navigate to **Spatial Adjustment | Adjust**. Note that the *Root Mean Square error* or RMS Error value is 4.9 meters. For a dataset with the reference scale 1:15,000, a RMS value of 4.9 meters is a reasonable one. The assumed reference scale and the resulting RMS value should be mentioned in a metadata file (**Item Description | Resource | Lineage**). Close the **Link Table**.

13. As you can see, the **Polyline** and **Polygon** layers were both transformed to match the Brussels-Orthophoto 2012 coordinate space. Inspect the results. On the **Editor Toolbar**, navigate to **Editor | Save Edits** to save the spatial adjustment applied to the Polyline and Polygon layers.

In the second part of this exercise, we will define the reference system for the Brussels_CADToGeodatabase feature dataset:

14. Open the **Catalog** window, and click on **Connect To Folder** `<drive>:\LearningArcGIS\Chapter5`. Expand World.gdb. Right-click on **Brussels_CADToGeodatabase**, and select **XY Coordinate System** in **Properties**. In the **Search** text box, type belge. Navigate to **Projected Coordinate Systems | National Grids | Europe** and select **Belge Lambert 1972**. Click on the **Add To Favorites** button. Click on **Apply**.

You will see a warning message which tells you that it cannot modify the existing object as it cannot acquire a schema lock. You cannot modify the reference system while you are in an open editing session.

15. Click on **OK** twice to close all dialog windows. On **Editor Toolbar**, select **Stop Editing** from the **Editor** menu. Repeat the last step to define the reference system for the Brussels_CADToGeodatabase feature dataset. Note that the Polyline and Polygon feature classes automatically inherited the feature dataset's reference system.

16. When finished, save your map document as MySpatialAdjustment.mxd to <drive>:\LearningArcGIS\Chapter5.

You can find the results of this exercise at <drive>:\LearningArcGIS\Chapter5\Results\SpatialAdjustment.

# Creating and editing attribute data

You can add a new attribute field to a layer table only outside of an edit session. When you add or delete a field in a feature class table, the file geodatabase structure (or schema) is updated. You cannot change the geodatabase schema inside the edit session. During the edit session, ArcGIS locks the schema of the feature class that is being edited, and the user can only create and edit feature shapes and edit the attribute values on the existing table structure.

# Editing the feature attributes

The process of attributes editing follows the same steps that are used to edit feature shapes.

There are two main ways to view and edit the feature attributes: using layer attribute table or using the **Attribute** dialog window.

In an open layer attribute table, the attribute values can be manually edited to a single cell or multiple selected cells. Using the **Field Calculator** tool, you can automate the editing process. You can perform simple or complex calculations on a field using mathematical and logical functions, in or out of an edit session. If you edit the attribute values with **Field Calculator** outside an edit session, you will not be able to undo the editing process.

The **Attribute** dialog window can be used only in an edit session. This displays the attribute values of the selected feature on the map.

Follow these steps to start editing feature attributes using the ArcMap application:

1. Open your map document called `MyEditingFeatures.mxd` from `<drive>:\LearningArcGIS\Chapter5\EditingFeatures`.

2. Set the `Buildings` layer as the only selectable layer in the map. From **Bookmarks**, select **Edit 1**.

3. In **Table Of Contents**, right-click on the **Buildings** layer and select **Open Attribute Table**. Dock the **Table** window to the bottom of the map display, as shown in the following screenshot:

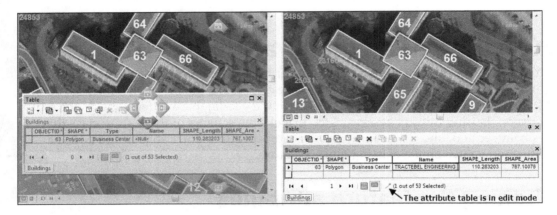

4. Inspect the attributes fields. Use the **Select Features** tool from the **Tool** toolbar to select the building feature with `ID = 63`, as shown in the previous screenshot. On the **Table** window, click on **Show selected records** to see only the selected record associated with the selected feature in the `Buildings` layer.

5. We would like to add a value in the **Name** field. First, we will start an edit session. If necessary, add **Editor Toolbar**. On **Editor Toolbar**, select the **Start Editing** option from the **Editor** menu.

6. In the **Table** window, note that the headings of the **Type** and **Name** fields have turned white, which means that these fields can be edited. Also, a small pencil has appeared at the bottom of the table window. This indicates that the `Buildings` attribute table is in edit mode.

7. If you click on the small gray box to the left of the selected record, the record and the corresponding feature on the map are changed to yellow. This helps you check which feature you are editing.

8. We will manually edit the building's name. Under the **Name** head field, click in the empty cell and type the name of building: `TRACTEBEL Engineering (GDF Suez)`. Click on the *Tab* key.

Next, we would like to edit more records at once using **Field Calculator**:

9. Click on the **Table Options** button to the top-left corner, and select **Select by Attributes**. Let's build the `Type = 'Business Center'` expression. Double-click on **Type** to add the field to the expression box. Click on the equal sign button (**=**). Click on the **Get Unique Values** button to display the unique values in the **Type** field. Double-click on the **Residential** values to add it to the expression box. Click on **Apply** and then click on **Close**.

10. In the **Table** window, check whether the **15 out of 53 Selected** text is displayed next to the pencil icon. Right-click on the **Type** field name and select **Field Calculator**.

11. In the **Field Calculator** window, type the text string enclosed with double quotes as follows: `"Business"`. Click on **OK**. The type of building has been changed for all fifteen features at the same time. Deselect the features on the map display using the **Clear Selected Features** tool from the **Tools** toolbar.

12. To save the attribute edits, navigate to **Editor | Save Edits** from **Editor Toolbar**.

In the second part of this exercise, we will change the name of five buildings using the **Attribute** tool, as shown in the following screenshot:

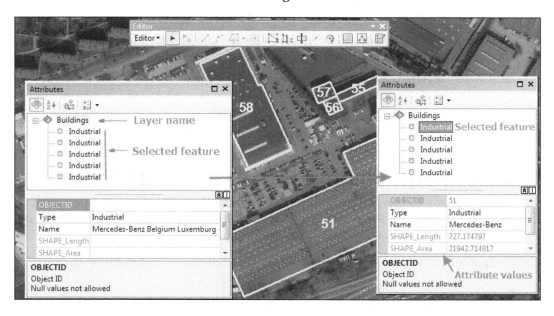

13. Close the **Table** window. From **Bookmarks**, select **Edit 2**.

14. Use the **Select Features** tool from the **Tool** toolbar to select five building features with ID 51, 55, 56, 57, and 58, as shown in the previous screenshot.

15. On the **Edit** toolbar, select the **Attributes** tool. The **Attributes** window displays the attributes values only for the selected building features.

16. Click on the first selected feature from the list. Note that the feature flashes on the map display and its corresponding attribute values are displayed at the bottom.

> If you want to see the feature IDs instead of its **Type** value, right-click on the **Buildings** layer in the **Attribute** window and navigate to **Layer Properties | Display**. Select the **OBJECTID** field from the **Field** drop-down list and click on **OK**. Stop the edit session, start it again, and select the features on the map to see your changes in the **Attribute** window.

17. Inspect the attribute values of all selected features. As all selected buildings have the same name, we would like to change the Name value from Mercedes-Benz Belgium Luxemburg to Mercedes-Benz for all five features at the same time. At the top of the **Attributes** window, click on the **Buildings** layer. The attributes associated with the **Buildings** layer are displayed.

18. Click in the cell next to the **Name** field and select the **Belgium Luxemburg** text. Right-click on the selected text and select **Delete**.

19. The shorter name should be updated for all five building features. Click on each selected feature from the list and inspect the value in the **Name** field.

20. Close the **Attribute** window. Deselect the features (**Tool | Clear Selected Features**) and save your edits.

21. Leave the MyEditingFeatures.mxd map document open to continue working with the Buildings attribute table in the next exercise.

# Creating and calculating an attribute field

In this section, we will learn how to add two new attribute fields in the Buildings attribute table. Next, we will estimate the cost of buildings based on their number of stories, SHAPE_Area values, and a fictive building price per square meter.

Follow these steps to add and calculate two new attribute fields for the Buildings layer:

1. In the MyEditingFeatures.mxd map document, right-click on the **Buildings** layer and select **Zoom To Layer** to see all buildings on the map display.

2. Open the **Buildings** attribute table. In the **Table** window, we will add two fields: `Levels` and `BuildingValue`. You cannot add a new field during an edit session. If necessary, save the previous edits and stop the edit session.

3. Click on the **Table Options** button at the top-left corner, and select **Add Field**. For **Name**, type `Levels`. For **Type**, accept `Short Integer`. Under **Field Properties**, click in the empty cell next to **Default Value** and type `1`. Click on **OK**. Open the **Add Field** window again to add a second field called `BuildingValue`. For **Type**, select `Long Integer` from the drop-down list.

4. To simplify this exercise, let's suppose that the residential buildings have four levels above ground (three stories) and the rest of building features have only one level above ground. Click on the **Table Options** button and then click on **Select By Attributes**. For the `Buildings` layer, build the `Type = 'Residential'` expression. Click on **Apply** and then click on **Close**. Click on **Show Selected Records** to see all nine records selected.

5. Next, you will edit the **Levels** attribute for the selected features. Right-click on the **Levels** field name and select **Field Calculator**. In the **Field Calculator** window, type `4`. Click on **OK**.

6. At the top of the table window, click on the **Switch Selection** button. For all 44 selected features, change the **Levels** value to `1` using the field calculator, as you did in the previous step. Click on **OK**.

7. Click on the **Show all records** button to display all the records in the `Buildings` table. Click on the **Clear Selection** button to deselect the records.

8. In the end, we will calculate the building values. Right-click on the **BuildingValue** field name and select **Field Calculator**. In the **Field Calculator** window, build the `[SHAPE_Area] * [Levels] *500` expression. The value, `500`, represents a fictive building cost per square meter in dollars. Click on the **Save** button and save your expression as `MyBuildingValue.cal` to `<drive>\LearningArcGIS\Chapter5`. Click on **OK**. ArcMap calculates the building value for each record and updates the `Buildings` attribute table.

9. Inspect the results in the **Table** window. If you want to have a more readable values in the **BuildingValue** field, right-click on it and select **Properties**. Under **Display**, click on the **Numeric** button and select **Show thousands separators**. Click on **OK** and then click on **Apply** to close the dialog windows.

10. You can use **Select By Attributes** in conjunction with **Field Calculator** to calculate the building values based on their **Type** values. For example, you can select all `Business` buildings and recalculate their values. Reuse the previous expression stored in `MyBuildingValue.cal` and change the building value per square meter to `1000` as follows: `[SHAPE_Area] * [Levels] *1000`.

11. Save your attribute edits. Stop the editing session. Save your map document as MyEditingAttributes.mxd to ..\Chapter5\EditingFeatures.

You can find the BuildingValue.cal expression and the resulting World.gdb geodatabase at ..\Chapter5\Results\EditingAttributes.

# Creating new features

So far, in this chapter, you worked with existing data. In this section, we will explore two ways of creating new features in a feature class: by digitizing and by $x$, $y$ coordinate pairs stored in a nonspatial table.

# Digitizing features

Creating new features using a base layer for reference is called onscreen or heads-up digitizing. For the background layer, you can use scanned and georeferenced paper maps, digital aerial photos, and satellite images.

There are three main aspects of the digitizing process: scale resolution of the reference layer, digitizing scale, and number of vertices in a feature's sketch.

The reference scale of your base layer influences the reference scale of your features. For example, if the pixel size of your orthophoto map (aerial photo) is 0.5 meter and if we consider that its planimetric precision corresponds to the scale 1:5,000, then the reference scale of your data may be assessed to 1:5,000. However, during the digitizing process the display scale of the orthophoto map should be larger than the scale 1:5,000. You should **Zoom in** on the base layer until you can distinguish the outlines of features clearly (for example, from 1:500 to 1:2,000). The sketch's vertices should accurately represent the shape of the real-world objects. A higher number of vertices don't increase the overall accuracy of your data, but they will significantly increase the size of your data.

Follow these steps to start digitizing features using the ArcMap application:

1. Open your map document called MyEditingFeatures.mxd from <drive>:\ LearningArcGIS\Chapter5\EditingFeatures.

2. Set the Buildings layer as the only selectable layer in the map. From **Bookmarks**, select **Edit 7**. At the top of **Table Of Contents**, click on the **List By Selection** button and make sure that the Buildings layer is the only selectable layer in the map.

3. The editing and digitizing processes follow the same editing steps. Thus, we will start an edit session. In **Table Of Contents**, right-click on the **Buildings** layer and navigate to **Edit Features | Start Editing**. The **Create Features** window opens and displays two feature templates corresponding to the layers available for editing.

4. In the **Create Features** window, double-click on the **Buildings** template to modify it. For **Description**, type Brussels buildings. For **Default Tool**, accept **Polygon**.

5. In this exercise, we will mainly digitize commercial buildings. In the empty cell next to the **Type** field, type Commercial. Click on **OK** to update the Buildings feature template. At the bottom of the **Created Features** window in the **Construction Tools** list, select the **Auto Complete Polygon** tool. Make sure that **Vertex Snapping** is active on the **Snapping** toolbar.

6. The **Auto Complete Polygon** tool allows you to create a new feature adjacent to on existing one. Move the pointer over the building corner (**Start** point). When the pointer snaps to **Buildings:Vertex**, click on it to add the first vertex (**Start**), as shown in the following screenshot:

7. Continue to add vertices on every building corner in a clockwise direction, as shown in the previous screenshot. If you want to delete a vertex during the sketch editing, right-click on vertex and select **Delete Vertex**. Add the last vertex on the **End** point and press the *F2* key to finish the sketch editing.

8. On the **Editor** toolbar, click on the **Attribute** tool. Note that the **Type** field has been populated with the default `Commercial` values defined in the feature template. Also, the **Levels** field is populated with the default value, `1`, defined in the previous exercise called *Creating and calculating an attribute field*. For **Name**, type `Mercedes-Benz`.

9. Save your edits and close the **Attribute** window. Deselect the building feature using the **Clear Selected Features** tool from the **Tools** toolbar.

10. Next, we will continue to create new building features. From **Bookmarks**, select **Edit 8**. In the **Create Features** window, check whether the `Buildings` template is still selected. Make sure that the default **Polygon** tool is selected in the **Construction Tools** list.

Next, we will digitize three buildings, as shown in the following screenshot:

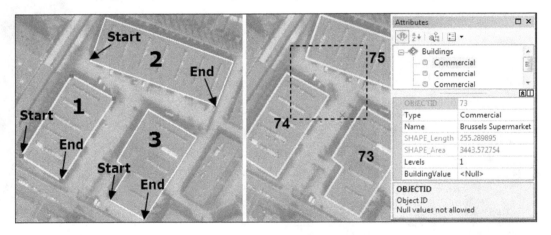

11. Let's add the first building. Click on the building corner (**Start**) to add the first sketch vertex. Add the next two vertices. Double-click on the last building corner (**End** point) to finish the sketch. Repeat this step to add the last two buildings, as shown in the previous screenshot.

12. With the **Edit** tool active, draw a box over all three building to select them, as shown in the previous screenshot.

13. On **Editor Toolbar**, click on the **Attribute** tool. Click on the **Buildings** layer, and type `Brussels Supermarket` in the cell next to the **Name** field. In the **Attribute** window, inspect the attribute values for all three selected features.

Let's create a multipart feature from these three selected building features:

14. Keep the **Attribute** window opened. From the **Editor** menu, select **Merge** from **Editor**. Select the first feature from the list and click on **OK**. Note that the **Attribute** window has been updated. There is a single selected feature which has SHAPE_Area of around 11300 square meters. On the map display, all three buildings have the same ID =70.

15. In the **Attribute** window, right-click on the feature and select **Open Table Showing Selection**. In the **Table** window, there is only one selected record with OBJECTID =70 corresponding to all three buildings on the map display.

16. On **Editor Toolbar**, click on the **Edit Vertices** tool to see the multipart sketch. To see or edit the *x*, *y* coordinate values of the sketch vertices, click on the **Sketch Properties** tool, as shown in the following screenshot:

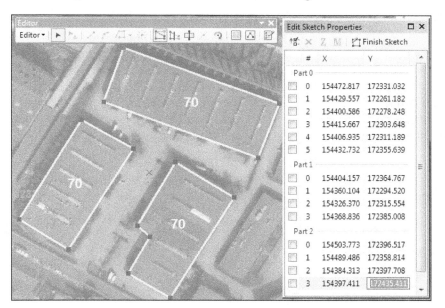

17. In the **Edit Sketch Properties** window, you can see the three sketch parts of the multipart polygon feature. To move the last vertex of the sketch, displayed in red on the map, click on its *y* coordinate value and type 172500. The vertex position should change on the map display. Press the *F2* key to finish the sketch editing. In the **Attribute** window, the SHAPE_Area value has automatically updated (increased to around 15100).

18. If you haven't save your edits, you can use the **Undo Merge** tool from the **Standard** toolbar to move the vertex back to its location and to separate the multipart feature into three individual polygon features.

If you have already saved your edits, use the **Explode Multipart Feature** tool from the **Advanced Editing** toolbar to separate a multipart feature into individual features.

19. Stop and save your feature edits. Save your map document as `MyDigitizingFeatures.mxd` to `..\Chapter5\EditingFeatures`.

You can find the results at `..\Chapter5\Results\ DigitizingFeatures`.

# Creating point features using XY data

In order to create new point features based on a table or a text file of tabular data, we need two fields that contain the X and Y coordinates separated by a delimiter (for example, comma or tab). Make sure that the field names do not contain spaces, dashes, or special characters.

The coordinate values may refer to any coordinate reference system using units, such as decimal degrees or meters. Using these two fields, ArcMap will create a point event layer temporarily stored only in the map document. Like any other layer, you can label the event features, change the symbology, set up the visible scale, query, display a subset of point events, and export the event layer to a permanent shapefile or feature class, like in the following screenshot.

In this subsection, we will create points for the Air Quality monitoring stations across the Europe using the *Air Quality e-Report* provided by the European Environment Agency, as shown in the following screenshot:

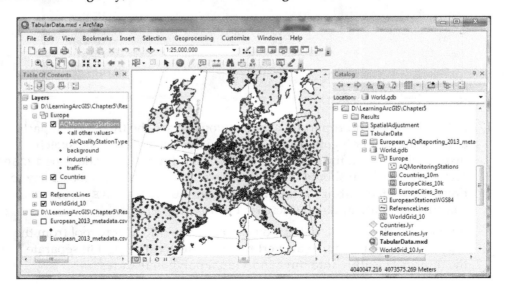

Follow these steps to create new features from tabular data using the
ArcMap application:

1. First, we will download a dataset from the European Environment Agency
   website. Open your internet browser and go to the URL, `http://www.`
   `eea.europa.eu/data-and-maps/data/aqereporting`. Make sure that
   the **European data** tab is selected. Under `European_AQeReporting_2013_`
   `metadata.zip (ZIP archive - 550KB)`, click on **Download file**. Save
   and unzip the archive to `<drive>:\LearningArcGIS\Chapter5`.

2. Start the ArcMap application and open a new map document. Open the
   **Catalog** window, and navigate to `<drive>:\LearningArcGIS\Chapter5`.
   Expand the `European_AQeReporting_2013_metadata` folder. Select the
   `European_2013_metadata.csv` table and drag it to the ArcMap map
   display area.

3. Note that the table has been added to **Table Of Contents** and the **List By
   Source** icon is active. The map display area is still empty.

4. Right-click on **Table** and select **Open**. Explore the attribute fields
   by scrolling through the table. The `Projection` field tells you that the
   coordinate reference system is `EPSG:4979` called `WGS84` (with World
   Geodetic System 1984 datum). The latitude, longitude, and ellipsoidal
   heights are stored in the next three fields. Close the table window.

5. Right-click on the table again, and select **Display XY Data**. A second option
   is to navigate to **Add Data | Add XY Data** from the **File** menu. The **X Field**
   and **Y Field** have already been detected. For **Z Field**, select **Altitude**.

6. Under **Coordinate System of Input Coordinates**, click on the **Edit** button.
   Navigate to **Geographic Coordinate Systems | World** and select **WGS 1984**.
   Click on **OK** to return to the **Display XY Data** window.

7. Click on **OK**. You will see a warning message that tells you that the table
   is missing the `ObjectID` field. As you will export the event layer to a point
   feature class in the file geodatabase, click on **OK**.

8. In **Table Of Contents**, a point event layer called **European_2013_metadata.csv
   Events** has been added. The point events represent the air quality monitoring
   stations in Europe.

9. Right-click on the **European_2013_metadata.csv Events** layer and select
   **Open Attribute Table**. Scroll to the end of the table, and note the **Shape**
   field, which stores the events geometry. Close the table window.

   As these point event layers are stored only within the map document, we
   will save the events layer as a point feature class stored within the `World.`
   `gdb` file geodatabase.

10. In **Table Of Contents**, right-click on point event layer by navigating to **Data | Export Data**. For **Output feature class**, click on the **Browse** button and navigate to `<drive>:\LearningArcGIS\Chapter5\World.gdb`. For **Name**, type `EuropeanStationsWGS84`. Click on **Save** and then click on **OK**.

11. Click on **Yes** to add the new standalone feature class as a layer. In **Table Of Contents**, right-click on the **EuropeanStationsWGS84** layer and navigate to **Properties | Source**. Inspect the information about its coordinate reference system. Click on **OK**.

12. Remove the table and point event layer in **Table Of Contents**.

13. To add the `EuropeanStationsWGS84` standalone feature class to the `Europe` feature dataset, you should first perform a CRS transformation from the `WGS84` geographic coordinate system to the `ETRS 1989 LAEA` projected coordinate system. Use the **Project** tool from **ArcToolbox**. Then, use the **Import** tool to add the projected feature class to the feature dataset.

14. We will use a shortcut using only the **Import** tool and accept the ArcMap suggested datum transformation from `WGS84` to `ETRS 1989 LAEA` (Lambert Azimuthal Equal Area).

15. In the **Catalog** window, right-click on the **Europe** feature dataset, and select **Import**. For **Input Feature**, select the **EuropeanStationsWGS84** layer. For **Output Feature Class**, type `AQMonitoringStations`. Under **Field Map**, erase the following fields: `Projection`, `Longitude`, `Latitude`, and `Altitude`. Click on **OK**.

16. To avoid storing duplicate data, remove `EuropeanStationsWGS84` from the `World.gdb` file geodatabase.

17. Change the `Layers` data frame's coordinate reference to `ETRS 1989 LAEA` (right-click on data frame, and navigate to **Properties | Coordinate System**) and change the **General | Display** units to `Meters`.

18. Select and drag the **AQMonitoringStations** feature class in the ArcMap map display area. Change the layer's default symbology by navigating to **Categories | Unique values | Value Field**, selecting `AirQualityStationType`, and removing the `<Null>` value from the legend list.

19. In addition to the `AQMonitoringStations` layer, you can drag the `Countries.lyr`, `ReferenceLine.lyr`, and `WorldGrid_10.lyr` layer files in the ArcMap display area for a good-looking final map.

20. Inspect the results. When finished, save your map document as `MyTabularData.mxd` to `<drive>:\LearningArcGIS\Chapter5`.

You can find the results of this exercise at `<drive>:\LearningArcGIS\Chapter5\Results\TabularData`.

# Summary

In this chapter, you identified the main steps in the editing process. You learned how to modify existing features and how to add new features. You explored different ArcMap instruments to edit the coincident features using **Auto-Complete Polygons** tool and using map topology tools.

You manually edited the attributes using the **Attribute** dialog window and the layer attribute table. You saw how to automate the attributes editing process using the **Field Calculator** instrument.

In the last part of this chapter, you explored another way to create a new feature class by importing a text file that contained $\varphi$, $\lambda$ real-world coordinates.

In the next chapter, you will learn how to use the ArcGIS analysis tools to obtain new information from your existing spatial datasets.

# 6
# Analyzing Geographic Data and Presenting the Results

In this chapter, we will explore one of the most important functions of a GIS: data analysis. We will visualize, prepare, and combine spatial datasets to obtain new information using the ArcGIS analysis tools. In the second part of this chapter, we will present the results of the spatial analysis through a professional-looking report.

By the end of this chapter, you will learn about the following topics:

- Planning for analysis and preparing data
- Associating spatial and nonspatial tables
- Using attribute and location queries
- Presenting the analysis results using tabular reports

## Planning the data analysis and preparing data

Throughout this book, you built a File Geodatabase, and created and edited different spatial data. During this chapter, we will plan, prepare, and perform a spatial analysis. There are seven main steps in the analysis process:

1. Define the problem or question.
2. Identify the conditions.
3. Identify the necessary data.
4. Plan the analysis process.
5. Prepare the data for analysis.

6. Perform the analysis and examine the results.

7. Present the results.

# Planning data analysis

In this section, we will identify the problem that has to be solved, the criteria that has to be used in spatial analysis, the spatial and non-spatial datasets that we will need, and the main steps of the actual analysis.

## Defining the problem

Let's suppose that a developer from Brussels plans to construct an apartment building in the Brussels-Capital Region. The scope of the analysis is to find the most suitable location for the new building. The local developer targets the young families with a medium income. The building should ideally be located near elementary and primary schools. The building should also be located close to sports or leisure facilities for children.

## Identifying the conditions

The location of the parcel that will be acquired to construct the apartment building should meet the following conditions:

- It should be in Brussels' municipalities with a population density greater than 20,000
- It should be on land without current use or discontinuous very low density urban fabric
- It should be within a municipality with high concentrations of young families ( <37 years old) with an income greater than 10,000 Euros.
- It should be within 500 meters of primary and secondary schools
- It should be at least 500 meters from fast transit roads and railways.

## Identifying the necessary data

The feature classes necessary to perform the analysis are the following:

- BrusselsCity: This feature class stores parcels; the parcels classified as fast transit roads and railways will be selected and exported in a new feature class named RoadsRailways
- Population density by municipalities: This is a hosted web layer from ArcGIS Online; this layer stores the Brussels' municipalities and data about the population

- `Average income per habitant by municipalities`: This is a hosted web layer from ArcGIS Online; this layer stores the municipalities and data about the population income

- `Brussels-Educational buildings`: This is a group layer from ArcGIS Online that is accessed directly over the web; this contains the `Primary schools` and `Secondary schools` layers

We will also need supplementary data about the population's average age for every Brussels' municipality from the `Brussels_AverageAge` nonspatial table, which is stored as a dBase (`.dbf`) format at `<drive>:\LearningArcGIS\Chapter6`.

It's necessary to investigate the metadata information of the datasets to find out about the data source, coordinate system, reference scale, and permission to use the data.

# Planning the analysis process

Planning the analysis helps you avoid mistakes along the analysis process. Throughout the planning step you may find that some additional layers or tables are necessary. It's a good idea to organize the main steps, data layers, and GIS tools into a workflow diagram, as shown in the following screenshot:

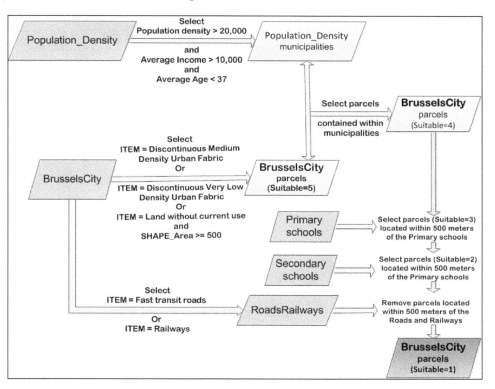

This workflow diagram identifies the attribute fields and values as derived from criteria mentioned in step two of the analysis process. In addition to this, this identifies the main GIS operations in preparing existing datasets and executing the analysis.

## Preparing data for analysis

This subsection refers to the fifth step of the spatial analysis process: *Prepare the data for analysis*. At this stage, you may need to define or change the coordinate systems of the datasets, add a new field to the attribute table, create table relationships, edit or update attributes or geometry of the features, and even correct the existing data errors.

In the next exercise, we will prepare the data layers mentioned in step three of the analysis process: *Identify the necessary data*.

First, we will add three different datasets from ArcGIS Online. Then, we will use the **Join** ArcMap method to attach more attribute fields to the attribute table of the `Population_Density` layer from the `Average_Income` attribute table and the `Brussels_AverageAge.dbf` nonspatial table.

The **Join** method connects two tables and virtually appends additional fields of one table (source) to another table (target) using a field that is common to both tables. These common fields or key fields do not need to have the same name, but they must be of the same field type and contain identical values (including capitalization and spaces). The table field types supported by ArcGIS for Desktop are: `Object ID`, `GUID`, `Text`, `Short Integer`, `Long Integer`, `Float`, and `Double`.

For example, the `Population_Density` attribute table and the `Brussels_AverageAge` table have a common field named `MUNC` and `Munc`, respectively. Both fields store the unique identifying code for the Brussels municipalities as the `Text` data type. The following screenshot shows a connection between the two tables (join) in a one-to-one relationship (called cardinality):

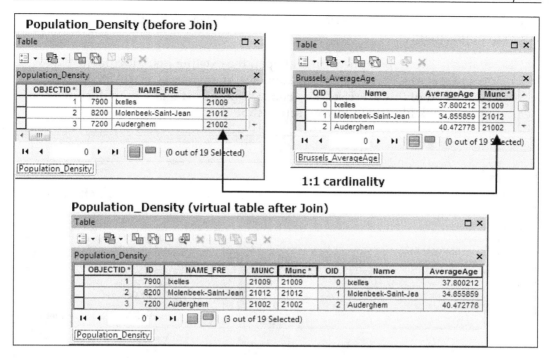

Cardinality defines how the individual record values from a table (target) relate to the individual record values from the second table (source). The **Join** connection works for one-to-one (*1:1*) and many-to-one (*M:1*) relationships. If you join two tables in one-to-many (*1:M*) and many-to-many (*M:M*) relationships, the resulting target table will omit all other records after the first value match of the each *target field value*. For the *M:M* and *1:M* relationships, use the **Relate** connection, which is the second method to associate two tables based on a common field in ArcMap.

For more information about tabular management, refer to ArcGIS Resource Center at http://resources.arcgis.com/en/help/.
Navigate to **Desktop (ArcMap): 10.4 | Manage Data | Data types | Tables | Joining and relating tables by attributes**.

The **Join** and **Relate** relationships between the table are saved in a map document (.mxd) as part of a layer or tables properties. Also, this means that they will be saved to a layer file (.lyr), layer package (.lpk), and map package (.mpk).

Follow these steps to start preparing data for the analysis using the
ArcMap application:

1.  Start the ArcMap application and open the existing map document
    `Brussels_Analysis.mxd` from `<drive>:\LearningArcGIS\Chapter6`.
    At the end of this exercise, the `Brussels_Analysis.mxd` map document
    should look like the following screenshot:

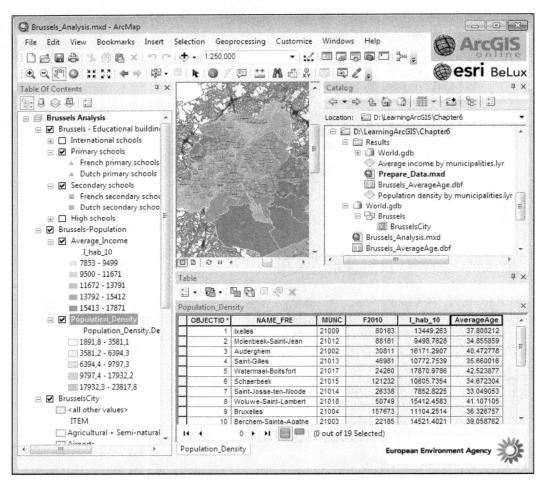

2. We will add three datasets for the Brussels city from ArcGIS Online. Check whether the ArcGIS for Desktop is connected to ArcGIS Online. On the **Standard** toolbar, click on the down arrow next to **Add Data,** and choose **Add Data from ArcGIS Online.**

3. Search for the `population brussels` data and select the **Brussels-Population density by municipalities 2013** feature service owned by `Esri_Belux_ Content`. Click on **Details** to read the information from the **Description** tab. Click on the yellow button called **Add Data.**

4. In **Table Of Contents,** right-click on the **Population density by municipalities** hosted web layer and select **Properties | Source.** Inspect the layer properties. The data source is a hosted feature service on ArcGIS Online. The pathname to the data source is the URL to the hosted web layers at `http://services1. arcgis.com/XLY7M6F9oPzjvKmJ/arcgis/rest/services/BXL_Communes_ Density/FeatureServer`. The `Population density by municipalities` web layer accesses the population density data that has been published by the `Esri_Belux_Content` owner and is hosted on ArcGIS Online.

5. Note that layer's coordinate system is `WGS_1984_Web_Mercator_Auxiliary_ Sphere`. Click on **OK.**

6. Next, we will export the data from the hosted feature layer named `Population density by municipalities` as a standalone feature class to the `World.gdb` file geodatabase. Right-click on the **Population density by municipalities** layer again and select **Export Data** from **Data.** For **Output feature class,** click on the **Browse** button and navigate to `<drive>:\ LearningArcGIS\Chapter6`.

7. Double-click on **World.gdb.** For **Name,** type `Population_Density` and select **File and Personal Geodatabase feature classes** from the **Save as type** drop-down list. Click on **Yes** to add data as a layer to the map.

8. To use the same symbols and classification classes as `Population density by municipalities` layer, right-click on the **Population_Density** layer and select **Symbology** from **Properties**. Click on the **Import** button in the top-right corner of the **Layer Properties** window; and for the **Layer** drop-down list, select the **Population density by municipalities** layer, as shown in the following screenshot:

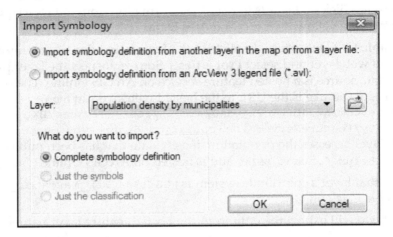

9. Click on **OK**, accept the default **Value Field** in **Import Symbology Matching Dialog**, and click on **OK** again.

You can use a different technique to symbolize the `Population_Density` layer:

Right-click on the **Population density by municipalities** layer and select the **Save As Layer File** option to create a layer file in the `<drive>:\LearningArcGIS\Chapter6` folder.

Add the layer file on the map and change its data source to the `Population_Density` feature class by navigating to **Layer Properties | Source**, and selecting **Set Data Source**.

10. In the **Layer Properties** window, click on **Apply**. Select the **Display** tab; and for the **Transparent** parameter, type `40`. Click on **OK** and inspect the results.

11. In **Table Of Contents**, right-click on the **Brussels-Population density by municipalities** layer and select **Remove**.

12. Repeat steps 2 to 10 to add the `Brussels-Average income per habitant by municipalities` dataset for Brussels city.

13. Click twice on the data frame name (**Layers**) and rename it `Brussels Analysis`. Right-click on the **Brussels Analysis** data frame and select **New Group Layer**. Click twice on the group layer and rename it `Brussels-Population`.

14. A group layer allows you to group and manage multiple layers. A group layer behaves like any other simple layer in **Table Of Contents**. Select the **Average_Income** and **Population_Density** layers and drag them on to the **Brussels-Population** group layer, as shown in the following screenshot:

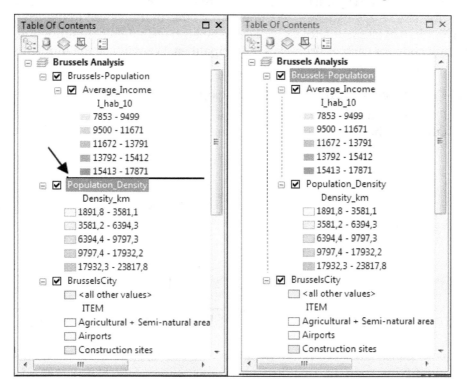

15. From ArcGIS Online, add the **Brussels-Educational buildings** hosted feature layer. In **Table Of Contents**, the **Brussels-Educational buildings** group layer contains four different layers. We will use only the `Primary schools` and `Secondary schools` layers. Instead of exporting the data from the web layer to the `World.gdb` file geodatabase, we will use the web layer in our GIS analysis by accessing it directly over the web. Turn off the `International schools` and `High schools` feature layers by unchecking them.

16. In **Table Of Contents**, right-click on the **Population_density** layer and select **Open Attribute Table**. Dock the **Table** window to the bottom of the map display.

17. Inspect the fields. The `Population_density` table contains the population values from 1980 to 2013. We also need more information about the population's income and age.

In order to gather the information in a single attribute table, we need to associate the `Population_Density` attribute table with the `Average_Income` attribute table and the `Brussels_AverageAge.dbf` nonspatial table that is stored in the `<drive>:\ LearningArcGIS\Chapter6` folder. All three tables have a field containing the unique identifying code for each of Brussels's municipalities. We will use the municipality's code values to join them:

18. Keep the `Population_Density` table open.

19. First, we will join the `Population_Density` and `Average_Income` attribute tables, as shown in the following screenshot:

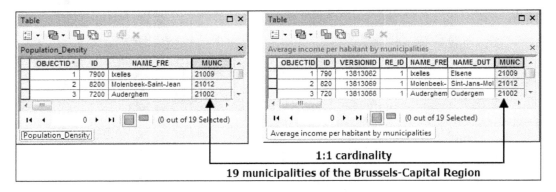

20. Both tables have a common field named **MUNC**, containing the unique identifying code for each Brussels municipalities. The field type is set to `String`.

21. Right-click on the **Population_density** layer and select **Join** from **Joins and Relates**. In the **Join Data** window, set the steps, as shown in the following screenshot:

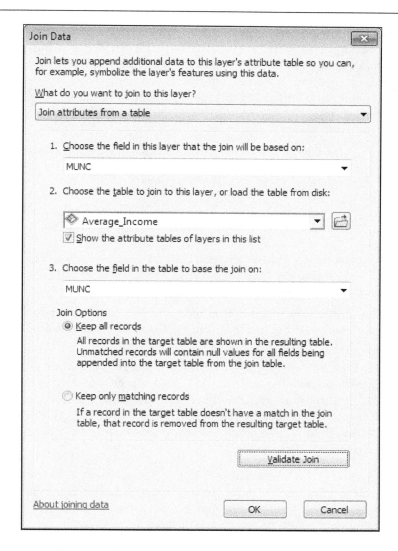

22. In the **Join Data** window, make sure the **Join attributes from a table** is selected. For step 1, select the **MUNC** field of the **Population_density** layer. For step 2, select **Average_income** as attribute table to join to the **Population_density** layer. For step 3, the **MUNC** field of the **Average_income** attribute table should be selected.

23. For **Join Options**, leave the **Keep all records** option selected. Click on the **Validate Join** button. The **Join Validation** window should return the following message: **All field and datasource validation tasks were completed successfully**. Click on **Close** and then **OK**.

24. Inspect the new added fields in the `Population_density` attribute table. These new appended fields from the `Average_income` attribute table do not affect the source data of the `Population_density` layer, and it is just a virtual connection in the current map document. You can remove the join by navigating to **Table Options | Joins and Relates** and clicking on **Remove Join(s)**.

 If you want to save this large table as a new permanent table in the `World.gdb` file geodatabase, you should export it by selecting **Export** from **Table Options**.

Next, we will join `Population_Density` and the `Brussels_AverageAge.dbf` nonspatial table:

25. Open the **Catalog** window, and navigate to `<drive>:\LearningArcGIS \ Chapter6`. Expand the `Chapter6` folder. Select the **Brussels_AverageAge** table and drag it to the ArcMap map display.

26. Note that **Table Of Contents** lists the layers by source in order to display the nonspatial table. At the top of **Table Of Contents**, the **List By Source** button is active.

27. In **Table Of Contents**, right-click on the **Brussels_AverageAge.dbf** table and select **Open**. Inspect the table fields. The **AverageAge** field stores information about the population's age in 2010. The **Code_Munc** field stores the unique identifying code for each of Brussels's municipalities. Right-click on the **Code_Munc** field name and choose **Properties**. Note the field type is set to `Short Integer`. Click on **OK**.

To join the `Population_Density` feature class and `Brussels_AverageAge` table, we need the common field to have the same data type. As the data type of the corresponding field in the `Population_Density` attribute table is set to `String`, we will add a new field to the `Brussels_AverageAge` table:

28. At the top of the `Brussels_AverageAge` table, click on the **Table Options** button and select **Add Field**. For **Name**, type `Munc`. For **Type**, choose `Text`. For **Length**, type `6`. Click on **OK**.

29. Next, we will populate the empty field with the values from the **Code_Munc** field. Right-click on the **Code_Munc** field name and choose **Field Calculator**. In the **Fields** scrolling list, double-click on the **Code_Munc** field to add it to the expression box. Click on **OK**. Note the new values for the **Munc** field.

30. To avoid storing duplicate data in the table, right-click on the **Code_Munc** field name, choose **Delete Field**, and click on **Yes** to confirm the deletion of the field.

Now, the `Brussels_AverageAge` table is ready to join the `Population_Density` attribute table:

31. In **Table Of Contents**, right-click on the **Population_density** layer and select **Join** from **Joins and Relates**. In the **Join Data** window, set the steps, as shown in the following screenshot:

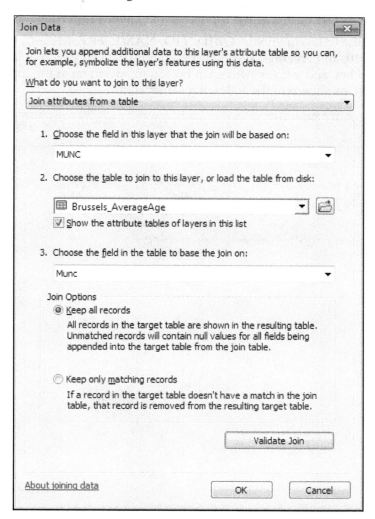

32. Click on **Validate Join** and then click on **Yes** on the **Create Index** dialog. Click on **Close** and then click on **OK** to join the tables.

33. In the **Table** window, close the **Brussels_AverageAge** table.

34. In the **Population_Density** attribute table, scroll across and look at the newly added fields. As we use only the information about the population's value, average income, and age, we will turn off all unnecessary and duplicate fields.

35. In **Table Of Contents**, right-click on the **Population_density** layer and select **Fields** from **Properties**. Uncheck all fields by clicking on the **Turn All Fields Off** button at the top-left of the **Layer Properties** window.

36. Check the following fields: **OBJECTID, NAME_FRE, MUNC, F2010, I_hab_10,** and **AverageAge**. Select all fields checked from the list using the *Ctrl* key and use the **Move Up** button to add them at the top of the field list, as shown in the following screenshot:

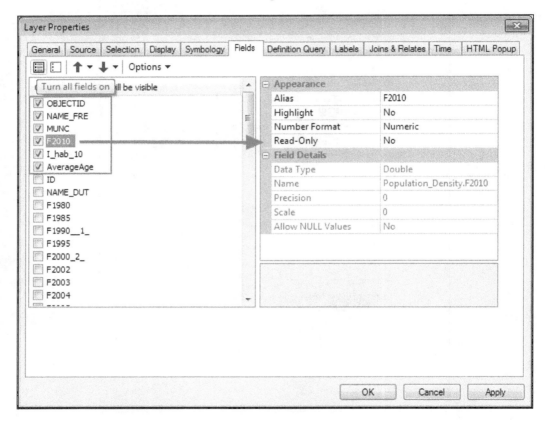

37. When finished, click on **OK** and inspect the changes in the Population_ Density table. Keep the table open.

38. In **Table Of Contents**, select the **List By Drawing Order** icon to no longer view the data sorted by source, as we finished working with the `Brussels_AverageAge` table.

39. When finished, save your changes in a new map document named `MyBrussels_Analysis.mxd` to `<drive>:\LearningArcGIS\Chapter6`.

40. Leave the map document open to start data analysis in the next section.

You can find the results at `<drive>:\LearningArcGIS\Chapter6\Results\PrepareData.mxd`.

# Performing the analysis

This section refers to the sixth step of the spatial analysis process: *Perform the analysis and examine the results*. After we have prepared the data, we are ready to follow the previous workflow diagram to perform the spatial analysis.

To identify the features that meet all analysis criteria, we will query the database. There are two types of queries in ArcGIS: *attribute query* and *location query*.

Attribute query helps us find features with particular attribute values; while location query identifies the features whose locations meet certain conditions that are relative to other features in the same layer or in a different layer.

We can combine the attribute and location queries during the analysis process to find features that meet your criteria. Working with the selected subsets of features has two types of approaches:

- **Additive**: This adds a new feature to the selected set of features.
- **Subtractive**: This removes features from the selected set of features.

# Attribute queries

To query a database based on attributes, a user has to build a *query expression*. A query expression has three basic components:

- Attribute field
- Operators: Logical (AND, OR, NOT), Relational (>=, <) and Arithmetic (+, -,%)
- Attribute value

Let's suppose that we want to know which parcels from the BrusselsCity layer are classified as fast transit roads. In ArcGIS, the first part of a standard SQL (Structured Query Language) expression is automatically created for us: **SELECT * FROM BrusselsCity WHERE**. As users, we should define the second part of the expression that represents the WHERE clause or criteria: ITEM = 'Fast transit roads and associated land'. The ITEM attribute field represents the *attribute field* that stores the classification information. The equal sign (=) represents the *operator*. The strings enclosed within single quotes represent the *attribute value*: 'Fast transit roads and associated land'.

Follow these steps to start performing the spatial analysis using the ArcMap application:

1. Open your map document called MyBrussels_Analysis.mxd from <drive>:\LearningArcGIS\Chapter6.

2. In the first part of this exercise, we will select all the fast transit roads and railways from the BrusselsCity layer.

3. In **Table Of Contents**, right-click on the **BrusselsCity** layer and select **Open Attribute Table**. In the **Table** window, click on the **Table Options** button and choose **Select By Attributes**.

4. In the **Select By Attributes** window, for **Method**, select **Create a new selection** from the drop-down list, as shown in the following screenshot:

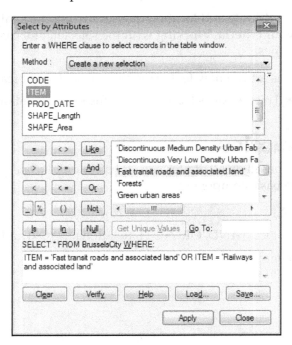

We will build a compound query expression that selects the features that are classified as roads or railways. A compound query uses an additional operator called **connector** (for example, OR, AND) to connect two or more different query expressions:

5. To build the query expression into the expression box, double-click on the **ITEM** field, and click on the equal to (=) button. Click on the **Get Unique Values** button and double-click on the **Fast transit roads and associated land** value in the **Unique Value** list.

 The strings are case sensitive and are always enclosed within single quotes.

6. To add the second criteria, click on the **Or** button to add the OR connector. The OR connector constrains that at least one of the two connected expressions must be true for the parcels to be selected.

7. Double-click on the **ITEM** field again and click on the equal to button. In the **Unique Value** list, double-click on **Railways and associated land value**.

8. Click on the **Verify** button to check the syntax. Save your query expression as SelectRoadRailways.exp at <drive>:\LearningArcGIS\Chapter6.

9. Click on **Apply** and then click on **Close**. Note that 140 features are selected on the map and their corresponding records are selected in the layer attribute table.

Next, we will save the set of selected features as a new feature class in the World. gdb\Brussels feature dataset:

10. In **Table Of Contents**, right-click on the BrusselsCity layer and select **Data | Export Data**. For **Output feature class**, click on the **Browse** button and navigate to ..\Chapter6\World.gdb.

11. Double-click on the Brussels feature dataset. For **Name**, type RoadsRailways, click on **Save**, and then click on **OK**. Click on **Yes** to add the exported data as a layer on the map.

12. In **Table Of Contents**, drag the **RoadsRailways** layer above the **Brussels-Population** group layer. Change the current symbol's color of the **RoadsRailways** layer to **Gray 60%**.

13. In the **Table** window, click on the **Clear Selection** button to deselect the records from the BrusselsCity attribute table.

14. Let's add a new field named `Suitable` that will store the parcel suitability values from 1 to 5. The value, 1, means the most suitable parcels. The value, 5, means the least suitable parcels. Click on **Table Options** and select **Add Field**. For **Name**, type `Suitable`. Set the field type to **Short Integer**. Click on **OK**.

Next, we will build another compound query expression that will select the municipalities that meet the following three criteria simultaneously: municipalities with a population greater than or equal to 20,000, with an average income greater than or equal to 10,000 Euro, and with an average age less than or equal to 37 years old:

15. At the bottom of the **Table** window, click on the **Population_Density** tab to see the attribute table of the `Population_Density` layer.

16. Click on the **Table Options** button and choose **Select By Attributes**.

17. In the **Select By Attributes** window, for **Method**, make sure that the **Create a new selection** option is selected.

18. Click on the **Load** button to add the query expression, as shown in the following screenshot:

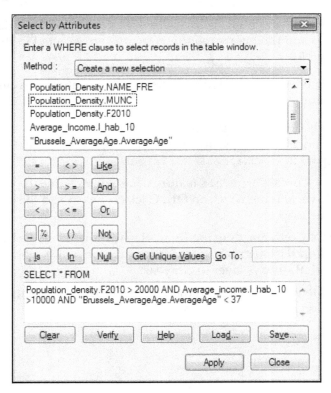

19. Navigate to `<drive>:\LearningArcGIS\Chapter6` and select the `Customers.exp` expression file. Click on **Open**. The AND connectors constrains that all three connected expressions must be true for the parcels to be selected.

20. In the **Select By Attribute** window, click on **Apply** and then click on **Close**. Look at the bottom of the `Population_Density` table to see that `3` records are selected.

Next, we will select all the parcels from the `BrusselsCity` layer that have an area greater than or equal to `500`, a medium to low density urban fabric, or are parcels without current use:

21. At the bottom of the **Table** window, click on the **BrusselsCity** tab to see the attribute table of the **BrusselsCity** layer.

22. Click on the **Table Options** button and choose **Select By Attributes**. In the **Select By Attributes** window, for **Method**, make sure that the **Create a new selection** option is selected.

23. Click on the **Load** button to add the query expression, as shown in the following screenshot:

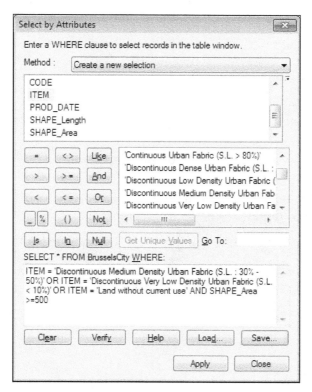

24. Navigate to `<drive>:\LearningArcGIS\Chapter6` and select the `SelectParcels.exp` expression file. Click on **Open**. In the **Select By Attribute** window, click on **Apply**, and then click on **Close**.

25. Click on the **Show selected records** button at the bottom of the `BrusselsCity` attribute table. There are 6686 records selected.

26. For the selected records, we will assign the suitability value 5. Right-click on the **Suitable** field name and select **Field Calculator**. Click inside the expression box and type 5. Click on **OK**. Keep the **Table** window open.

27. On the **Standard** toolbar, click on the **Save** button to save your changes in the current map document. Leave the map document open to continue the analysis in the next section.

You can find the `SelectRoadRailways.exp` expression file and the partial results in the `Brussels_Analysis_SelectByAttributes.mxd` map document at `<drive>:\LearningArcGIS\Chapter6\Results`.

# Location queries

There are four types of spatial relationships that you can explore with the location query: *intersect by*, *within a distance of*, *inside of*, and *adjacent to*.

For example, we would like to select parcels from the `BrusselsCity` layer that are completely within a specified municipality of Brussels city. To query a database based on location, users have to build a location query that consists of the following three things:

- The layer that contains the parcels that we want to select (the `BrusselsCity` layer)

- The layer that contains related municipality features (the `Population_Density` layer)

- The location relationship or the spatial selection method (`Completely contain the source layer feature`)

Follow these steps to continue performing the GIS analysis using the location queries:

1. Open your map document called `MyBrussels_Analysis.mxd` from `<drive>:\LearningArcGIS\Chapter6`.

First, we will reduce the currently selected parcels from the `BrusselsCity` layer to only the ones that are located within the three selected municipalities in the `Population_Density` layer:

1. From the **Selection** menu, click on **Select By Location** to open the dialog window, as shown in the following screenshot:

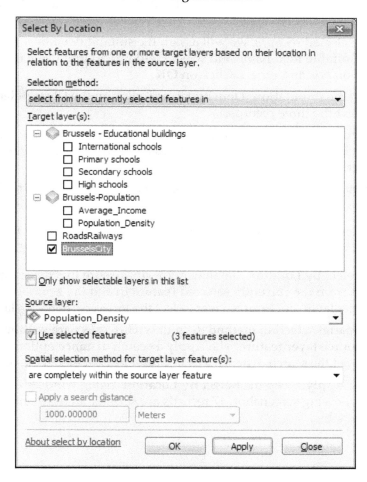

2. In the **Selection method** drop-down list, choose **Select from the currently selected feature in**. For **Target layer**, check the **BrusselsCity** layer. For **Source layer**, select the **Population_Density** layer. Select **Use selected features**.

3. In the **Spatial selection method** drop-down list, select **are completely within the source layer feature**. Click on **Apply**. Keep the **Select By Location** dialog window open. Move the **Select By Location** dialog window to the right-hand side of the data frame if necessary so that you can see the map and the **Table** window.

4. In the **Table** window, look at the bottom of the BrusselsCity attribute table to see that 65 records are selected.

5. For the selected records, we will assign the suitability value 4. Right-click on the **Suitable** field name and select **Field Calculator**. Click inside the expression box and type 4. Click on **OK**.

6. Click on the **Population_Density** tab and click on the **Clear Selection** button to deselect the three records.

7. Now, on the map, only the 65 parcels from the BrusselsCity layer that meet the previous specified criteria are selected. Return to the BrusselsCity table by selecting its corresponding tab.

8. Save your changes in the current map document using the *Ctrl+S* keys combination.

Next, we will refine the selected parcels using another criterion: only those parcels that are located within 500 meters of Primary and Secondary schools:

9. In the **Select By Location** dialog window, make sure that the selection method is **Select from the currently selected feature in** and the BrusselsCity layer is the **Target layer**. For **Source layer**, select the Primary schools layer.

10. In the **Spatial selection method** drop-down list, select **are within a distance of the source layer feature**. The **Apply a search distance** option is selected by default. Set the search distance to 500 and select Meters in the drop-down list.

11. Click on **Apply**. Keep the **Select By Location** dialog window open. As shown in the following screenshot, 47 records are selected in the BrusselsCity attribute table:

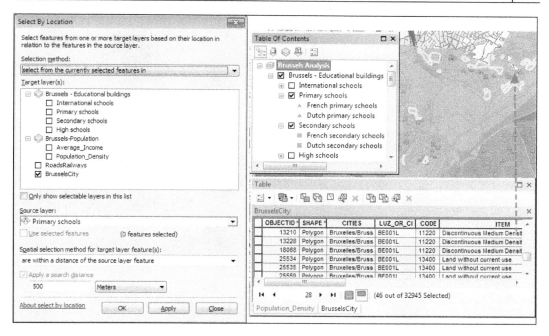

12. For the selected records, we will assign the suitability value 3. Right-click on the **Suitable** field name and select **Field Calculator**. In the expression box, type 3. Click on **OK**.

13. In the **Select By Location** dialog window, make sure that the **BrusselsCity** layer is the **Target layer(s)**. For the **Source layer**, select the **Secondary schools** layer.

14. Make sure that the selection method is **are within a distance of the source layer feature** and that the search distance is 500 Meters. Click on **Apply** and then click on **Close**.

15. In the **Table** window, there are 20 selected records in the BrusselsCity attribute table. For these selected records, assign the suitability value 2 in the Suitable attribute field.

16. Save your changes in the current map document.

In the last part of this exercise, we will find the parcels that meet the last criteria: the parcels should be at least 500 meters away from the fast transit roads and railways:

17. In the **Select By Location** dialog window, set the parameters, as shown in the following screenshot:

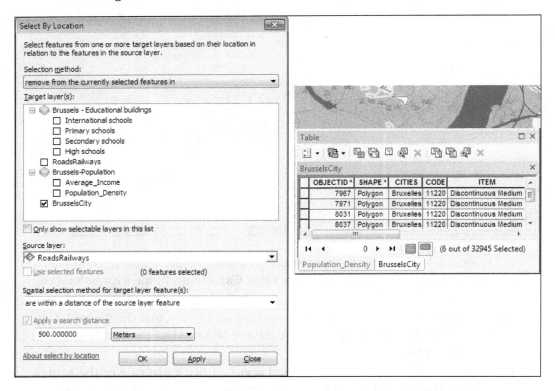

18. In the **Selection method** drop-down list, choose **remove from the currently selected feature in**. Make sure that the **BrusselsCity** layer is the **Target layer**.

19. For the **Source layer**, select the RoadsRailways layer. In the **Spatial selection method** drop-down list, select **are within a distance of the source layer feature**. For the **Apply a search distance** option, set the search distance to 500 Meters.

20. Click on **Apply** and then click on **Close**. As shown in the previous screenshot, six records are selected in the BrusselsCity attribute table. The parcels that were removed from the previous selected set of parcels are located too close to the fast transit roads and railways.

21. For all six selected records, assign the suitability value 1 in the **Suitable** attribute field. Inspect the attributes for all these six parcels that meet all of the criteria of our scenario.

You can export the selected set of parcels as a new feature class to `World.gdb` to use it in other map documents or as data input for further GIS analysis.

22. Keep the **Table** window open and the set of parcels selected on the map. Save the `BrusselsCity` layer as `Layer File` and name it `AnalysisResults.lyr`. This layer file references the data source and keeps the parcels selected. We will use this layer later on.

23. On the **Standard** toolbar, click on the **Save** button to save your changes in the current map document.

24. Leave the map document open to start creating a report to present your results in the next section.

You can find the results at `<drive>:\LearningArcGIS\Chapter6\Results\ Brussels_Analysis_SelectByLocation.mxd`.

# Creating a data report

So far, in this chapter, you performed an analysis of the most suitable parcels for a new residential building in the Brussels Capital region. In this section, we will create an analysis report for the local developer who wants to examine the analysis results and take the best decision for his construction project.

A report allows you to organize, format, print, and present the tabular data. In a report, you can make the following customizations:

- Select the attribute fields that will be displayed
- Display all records or the subsets of records in a table
- Sort the records based on their values (for example, sort the parcels by their area value)
- Calculate summary statistics (for example, average and standard deviation)
- Include logo, images, or page numbers
- Set the tabular format (for example, a classic columnar format)

When you finish organizing and formatting the report, you can save it as a Report Layout File (`*rlf`). This format will allow you to edit and update the report in ArcGIS for Desktop. You can also export the report to other file formats, such as Portable Document Format (PDF), TIFF format (TIF), or even Microsoft Excel (XLS).

Follow these steps to start creating a report of your analysis results using the ArcMap application:

1.  Open your map document called `MyBrussels_Analysis.mxd` from `<drive>:\LearningArcGIS\Chapter6`.

2.  From the **Bookmarks** menu, select **Manage Bookmarks**. Click on the **Load** button and navigate to `<drive>:\LearningArcGIS\Chapter6`. Select the `Report.dat ArcGIS Place` file and click on **Open**. From the **Bookmarks** menu, select the **Report** bookmark.

3.  If necessary, in the **Table** window, use **Select By Attribute** to select all six parcels that meet the `Suitable = 1` criteria.

4.  Let's create a picture of the current extent. From the **File** menu, select **Export Map**. For **Save in**, navigate to `<drive>:\LearningArcGIS\Chapter6`. From the **Save as type** drop-down list, select the `JPEG` format. For **File name**, type `MyBrussels_Analysis_Report`. Click on **Save**. We will use this picture later on in this exercise.

Next, we will change the properties of the `SHAPE_Area` field to improve the value display on the report:

5.  Right-click on the **SHAPE_Area** field name and select **Properties**. For **Alias**, type `Area (square meters)`. Click on the ellipsis button next to **Numeric**. Change the number of decimal places to `0`. Select **Show thousands separators**. Click on **OK** twice to close the dialog windows.

6.  Inspect the results in the **Table** window.

Now, the selected records are ready to add to the report:

7.  Click on **Table Options**, and select **Create Report** in **Reports**. In the **Report Wizard** dialog window, select **BrusselsCity** from the **Layer/Table** drop-down list. Add the **OBJECTID, CODE, Area (square meters)**, and **ITEM** fields in the **Report Fields** list, as shown in the following screenshot:

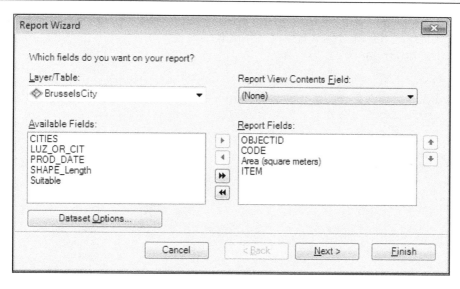

8.  Click on **Dataset Options,** and select **Selected Set.** Click on **OK** and then click on **Next.** As we don't need to group the fields for the report, accept the default settings and click on **Next.**

9.  We will sort parcels by area value. Under **Fields,** select the **Area** (square meters) field. Under **Sort,** select **Ascending.**

10. Click on **Summary Options** and select the **Avg, Max,** and **Min** values. For the **Available** sections, select **End of Report.** Click on **OK** and then click on **Next.**

11. Accept the default layout and style for the report in the next two panels. For the report title, type `Best parcels in Brussels City`. Select the **Preview the report** option and click on **Finish.**

12. The **Report Viewer** window opens and displays the report.

In the next steps, we will customize the appearance of the report elements, such as the spacing of the field names, the alignment of text, the value display, and the background of the report:

13. At the top-left side of the **Report Viewer** window, click on the **Edit** tool. First, we will increase the space between the field names in the **pageHeader** section, as shown in the following screenshot:

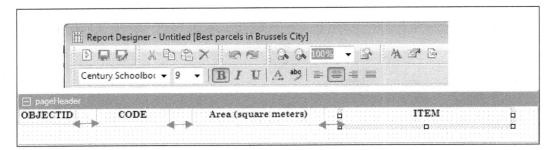

14. In the **pageHeader** section, click on the ITEM field name. Drag it to the right-hand side of the **pageHeader** section. Repeat this step for the Area and CODE fields. Make sure that there is enough space between the field names.

15. Next, select all field names using the *Shift* key and click on each field name again so that all four fields are selected. Then, click on the **Bold** and **Align Center** buttons.

16. At the top-left side of the **Report Designer** window, click on the **Run Report** button to preview our changes on **pageHeader**. As you can see in the **Report Viewer** window, the report will look much better if we have more space between the field headers and the values. Also, the data should be aligned with the field names.

17. Click on the **Edit** tool to return to the **Report Designer** dialog window.

18. Click on the line between the **pageHeader** and **detail** sections. When the pointer turns into a small two-headed arrow, drag down to increase the **pageHeader** width, as shown in the following screenshot:

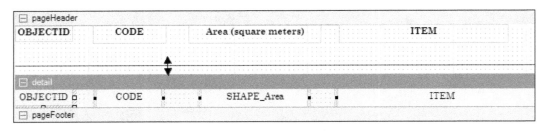

19. Under the **detail** section, select the ITEM element and drag it under its corresponding field name from the **pageHeader** section. Repeat this step for the rest of the data elements in the **detail** section.

20. When finished, select all four data elements and click on the **Bold** and **Align Center** buttons. With all four data elements selected, in the **Element Properties** section, set **Back Color** in **Appearance** to **Transparent**, as shown in the following screenshot:

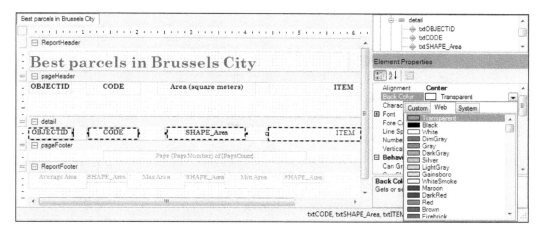

21. Under **ReportFooter**, select **Average Area**. In the **Element Properties** section, scroll down to see **Data | Text**. Change the text to **Average Area**. Repeat this step for the **Max Area** and **Min Area** summary values.

22. Preview the changes using the **Run Report** button. The report formatting should look similar to the following screenshot:

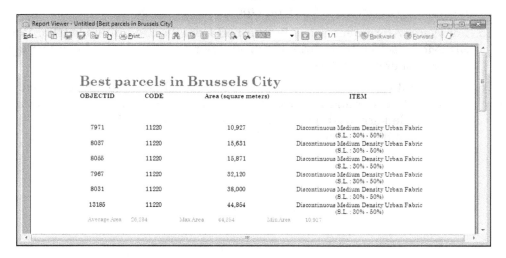

In the last part of this exercise, we will display the map picture as a watermark on the report:

23. At the upper-right of the **Report Designer** window, click on **Report**, as shown in the following screenshot:

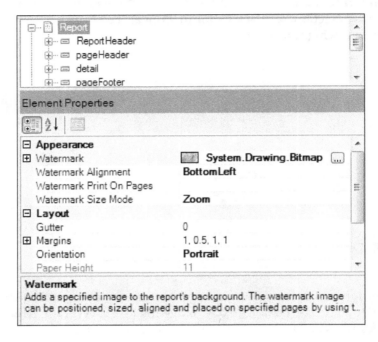

24. In the **Element Properties** section, click on **Watermark**. Click on the ellipsis button next to **Watermark**, navigate to ..\Chapter6, and select MyBrussels_ Analysis_Report.jpg. Click on **Open**.

25. For **Water Alignment**, select **BottomLeft** from the drop-down list. For **Water SizeMode**, select **Zoom** from the drop-down list. Accept the default values for the rest of parameters.

26. If you want to add the logo of the data sources, click on the line between the **ReportHeader** and **pageHeader** sections. When the pointer turns into a two-headed arrow, drag down to increase the **ReportHeader** width, as shown in the following screenshot:

27. Select the report title and drag it to the bottom of the **ReportHeader** section, as shown in the previous screenshot.

28. On the left-hand side of the **Report Designer** window, select the **Picture** element under **Design Elements**. Draw a rectangle above the report title. In the **Elements Properties** section, click on the ellipsis button next to **Image Source**.

29. In the **Image Source** window, select **Use a static image**. Click on the **Browse** button, navigate to ..\Chapter6\Logo, and select the arcgisonline.jpg image. Click on **Open** and then click on **OK**. Repeat these steps for the last two logos: Esri BeLux and European Environment Agency (EEA).

30. Preview the changes in **Report Viewer** using the **Run Report** button. The report formatting should look similar to the report named BestParcels_BrusselsCity.pdf that is stored at ..\Chapter6\Results.

31. We will save the report as a report layout file that allows you to use it again in ArcGIS for Desktop. At the upper-right of the **Report Viewer** window, click on the **Save as** button. Save your report as MyBestParcels_BrusselsCity.rlf in the Chapter6 folder.

32. If you want to save the report as a PDF file, click on the **Export report to file** button. For **Export Format**, select **Portable Document Format** (PDF). For **Filename**, click on the ellipsis button.

33. Navigate to ..\LearningArcGIS\Chapter6. For the **File** name, type MyBestParcels_BrusselsCity. Click on **Save** and then click on **OK**.

34. Close **Report Viewer**. If you want to open the MyBestParcels_BrusselsCity.rlf report in ArcMap once again, from the **View** menu, choose **Load Report** in **Report** and navigate to your report.

If you want to open an existing report file in a new map document, you must specify the layout or data source. Try to open your report in a new empty map document. You have to specify `AnalysisResults.lyr` as the layer source. The `MyBestParcels_BrusselsCity.rlf` report uses the selected parcels from the `BrusselsCity` layer. For this reason, we have saved these changes as a new layer file named `AnalysisResults.lyr` in the last part of the previous exercise.

35. Save your map document as `MyBrussels_Analysis_Report.mxd` to `..\LearningArcGIS\Chapter6`.

You can find the `BestParcels_BrusselsCity.rlf` report at: `<drive>:\LearningArcGIS\Chapter6\Results`.

# Summary

In this chapter, you got acquainted with the geographic analysis. You experienced the main steps in the spatial analysis process. Throughout the multistep GIS analysis, you combined the attribute and location queries to find a specific set of features based on their attributes and locations.

In the next chapter, you will continue to perform spatial analysis using the geoprocessing tools, which are an integral part of the ArcGIS analysis processes.

# 7

# Working with Geoprocessing Tools and ModelBuilder

In this chapter, we will learn how to combine the existing spatial datasets to create new datasets using data extraction, proximity, and overlay analysis tools. Also, we will build a geoprocessing workflow to perform a complex GIS analysis using the ModelBuilder graphical environment.

By the end of this chapter, you will learn about the following topics:

- Working with proximity and overlay analysis tools
- Creating geoprocessing models using ModelBuilder

## Working with Geoprocessing tools

The Esri GIS Dictionary defines **geoprocessing** as: "a GIS operation used to manipulate GIS data." A geoprocessing tool allows you to perform a single and specific geoprocessing operation using at least one dataset to produce a new dataset. Most of the geoprocessing tools create new datasets, leaving the existing datasets intact. In this context, geoprocessing occurs during the following:

- Copying, importing, and adding features from one dataset to another dataset (for example, the **Feature To Point** system tool from **ArcToolbox**, the **Data Management** toolbox, the **Features** toolset)
- Converting datasets (for example, the **CAD to Geodatabase** system tool from the **Conversion Tools** toolbox, the **To Geodatabase** toolset)
- Analyzing data using specific geoprocessing operations (for example, the **Clip** tool from the **Analysis Tools** toolbox, the **Extract** toolset)

There are three main categories of geoprocessing tools for spatial data analysis in **ArcToolbox**:

1.  **Extraction**: This creates a subset of features that are based on a specified extent
2.  **Overlay**: This combines features and their attributes
3.  **Proximity**: This locates features based on their distance from other features, and computes the distance between features

Through a geoprocessing operation, a geoprocessing tool works with the following main parameters:

*   The name and location of the input dataset
*   The name and location of the output dataset
*   One or more values that are specific to the operation performed (such as the distance value for the **Buffer** tool)

In this chapter, we will plan, prepare, and perform a site selection analysis. We will go through the seven steps in the analysis process mentioned in *Chapter 6, Analyzing Geographic Data and Presenting the Results*.

# Planning data analysis

In this subsection, we will define the problem, the criteria used in the spatial analysis, the necessary datasets, and the workflow diagram. Let's suppose that the city councilors of the city of Brussels plan to construct a pediatric care center to provide medical services to children in primary and secondary schools. The aim of the analysis is to find the most suitable location for the new building.

The location of the parcel that will be acquired to construct the pediatric center should meet the following conditions:

*   It should be in the Brussels' municipality with a population density greater than 10,000 with, and more than 8 schools per square kilometer
*   It should be land without current use
*   It should be within 300 meters of primary and secondary schools
*   It should be at least 600 meters away from existing hospitals

The feature classes that are necessary to perform the site selection analysis are the following:

*   The BrusselsCity feature class from the European Environment Agency (http://www.eea.europa.eu/data-and-maps/data/urban-atlas/)

- The Population density by municipalities and Hospitals web layers that are published by the Esri_Belux_Content owner and hosted on ArcGIS Online

- The Brussels-Educational buildings group layer that is published by the Esri_Belux_Content owner and hosted on ArcGIS Online; this contains the Primary schools and Secondary schools layers

The coordinate reference system of the dataset is Belge_Lambert_1972 with the Lambert_Conformal_Conic map projection and the D_Belge_1972 datum.

The workflow diagram displays the following tasks:

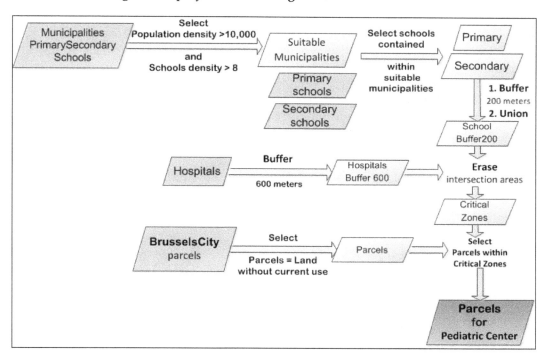

In the workflow diagram you may see the following tasks described as:

1. Identifying the Brussels' municipalities with a population density of over 10,000 inhabitants and with more than 8 schools per square kilometer using the **Spatial Join** and **Select** tools.

2. Identifying the schools within the municipalities selected in the first step using the **Clip** tool.

3. Creating a 300 meters buffer around each school within the suitable municipalities and a 600 meters buffer around each hospital using the **Buffer** tool.

4. Identifying critical areas that are not close to a hospital and need medical services using the **Erase** tool.

5. Identifying parcels without current use and which intersect with the critical areas using the **Select** and **Select Layer By Location** tools.

6. Saving the suitable parcels to build a pediatric care center in a new feature class called Parcels_PediatricCenter using the **Copy Features** tool.

# Preparing data for analysis

This subsection refers to the fifth step of the spatial analysis process: *Prepare the data for analysis*, covered in *Chapter 6, Analyzing Geographic Data and Presenting the Results*. Before you start running a geoprocessing tool, set the main parameters of **Environment Settings**, such as the output geodatabase for all derived datasets and output coordinate systems. The geoprocessing environment settings influence the results of the tools. The following screenshot shows the hierarchy of **Environment Settings**:

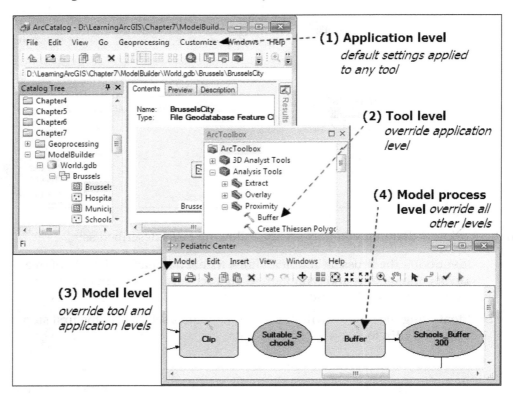

For more information about the geoprocessing tools and the **Environment Settings**, please refer to ArcGIS Resource Centre: http://resources.arcgis.com/en/help/.

- Navigate to **Desktop (ArcMap): 10.4 | Analyze | Commonly used tools**
- Navigate to **Desktop (ArcMap): 10.4 | Tools | Environments**

Follow these steps to start preparing data for the site selection using the ArcMap application:

1.  Start the ArcMap application and open the existing Site_selection.mxd map document from <drive>:\LearningArcGIS\Chapter7\Geoprocessing. The map document contains layers that represent the parcels, municipalities, hospitals, and primary and secondary schools in Brussels.

2.  Open the **Catalog** window and connect to the Chapter7 folder. Click on the **Connect To Folder** button, navigate to <drive>:\LearningArcGIS\Chapter7, and click on **OK**.

3.  Expand the Chapter7\Geoprocessing\World.gdb file geodatabase. In this exercise, we will work with the feature classes stored in the Brussels feature dataset. Expand it to see its contents.

4.  Before we start performing any geoprocessing task, we should set the main ArcMap environment parameters. From the **Geoprocessing** menu, click on **Environments**. In the **Environment Settings** window, expand the **Workspace** category. Click **Show Help** to see the meaning of every parameter.

5.  We will store all output feature classes resulting from the geoprocessing analysis to the Brussels feature dataset. Set **Current Workspace** and **Scratch Workspace**, as shown in the following screenshot:

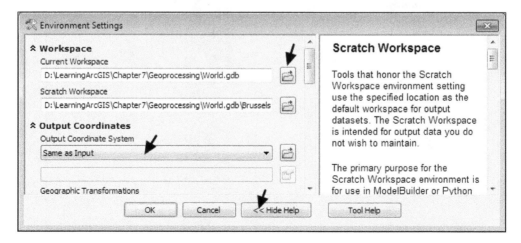

6. Expand the **Output Coordinates** category and make sure that **Output Coordinate System** is set to **Same as Input**. Click on **OK**.

Next, we will use a *point in polygon overlay operation* to count the number of primary and secondary schools in the Brussels municipalities:

7. From the **Geoprocessing** menu, click on **Search For Tools**. Dock the **Search** window on the right-hand side of the ArcMap window. In the **Search** window, make sure that **Local Search** is selected. In the keyword box, type `spatial join` and click on the blue icon next to the search box.

8. In the **Search** window, click on the **Spatial Join (Analysis)** tool.

In the **Search** window, click on the **Auto Hide** button to make the window point down.

If you want to read the **Spatial Join (Analysis)** documentation before you start running it, click on the descriptive text below the blue item name to open the **Item Description** window.

9. In the **Spatial Join** dialog window, click **Show Help** to see the meaning of every parameter. Set the tool's parameters, as shown in the following screenshot:

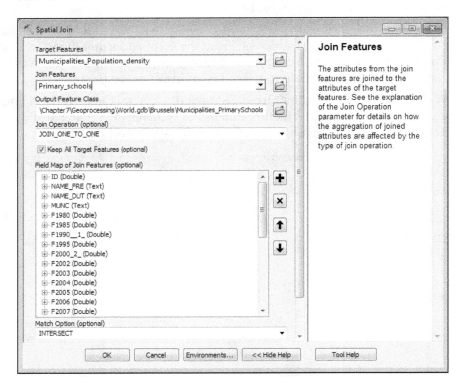

10. For **Target Features**, select **Municipalities_Population_density** layer. For **Join Features**, select the **Primary_schools** layer. Note that **Output Feature Class** is already set at `Chapter7\Geoprocessing\World.gdb\Brussels` with a default name. Change the feature class name to `Municipalities_PrimarySchools`. Make sure that **Join Operation** is set to `JOIN_ONE_TO_ONE` and **Match Option** is set to **INTERSECT**.

11. Click on **OK** to run the tool. When finished, select **Close this dialog when completed successfully** and click on **Close**. The output layer is added in **Table Of Contents**. Open the attribute table for the `Municipalities_PrimarySchools` layer. Note the `Join_Count` field that stores the number of primary schools in every Brussels municipality. Keep the **Table** window open.

Next, we will use the **Spatial Join** tool again to count the number of secondary schools in the Brussels municipalities:

12. Use the **Search** window to find and open the **Spatial Join** tool. Set the following parameters:

    ° **Target Features**: Set this to `Municipalities_PrimarySchools`

    ° **Join Features**: Set this to `Secondary_schools`

    ° **Output Feature Class**: Set this to `..World.gdb\Brussels\Municipalities_PrimarySecondarySchools`

    ° **Join Operation (optional)**: Set this to `JOIN_ONE_TO_ONE`

    ° **Match Option (optional)**: Set this to `INTERSECT`

13. Accept the default values for all other parameters. Click on **OK** to run the tool.

14. When finished, the output layer is added in **Table Of Contents**. Open the attribute table for the `Municipalities_Population_density` and `Municipalities_PrimarySecondarySchools` layers. Arrange the tables using the blue docking arrows so that you can see all of them at the same time, as shown in the following screenshot:

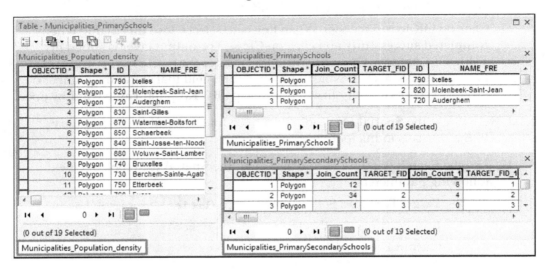

15. Inspect the tables. In the `Municipalities_PrimarySecondarySchools` table, the **Join_Count** and **Join_Count_1** fields store the number of primary and secondary schools in Brussels municipalities.

16. Next, we will calculate the schools density for every municipality. Add a new field in the `Municipalities_PrimarySecondarySchools` table. Click on the **Table Options** button and select **Add Field**. For **Name**, type `Schools_ Density`. For **Type**, choose `Short Integer`. Click on **OK**.

17. Right-click the **Schools_Density** field name and choose **Field Calculator**. Click on the **Load** button and select the `Schools_Density.cal` expression file from `<drive>:\LearningArcGIS\Chapter7\Geoprocessing`. Inspect the expression and click on **OK**.

18. When finished, save your changes in a new map document named `MySite_ Selection.mxd` to `<drive>:\LearningArcGIS\Chapter7\Geoprocessing`.

19. Leave the map document open to start data analysis in the next section.

You can find the results at `<drive>:\LearningArcGIS\Chapter7\Results\ Prepare_Data.mxd`.

# Performing site selection

This subsection refers to the sixth step of the spatial analysis process: *Perform the analysis and examine the results* covered in *Chapter 6, Analyzing Geographic Data and Presenting the Results.*

We will perform a site selection, which is the most common type of spatial analysis, to answer the following question: *What are the best locations for a Pediatric Care Center in the City of Brussels?*

To identify the most suitable parcels from the `BrusselsCity` feature class, we will use the spatial criteria already defined in the previous subsection named *Planning Data Analysis.*

Follow these steps to perform a site selection analysis using the ArcMap application:

1. Start the ArcMap application and open your map document named `MySite_selection.mxd` from `..\Chapter7\Geoprocessing`.

First, we will select municipalities with a population density greater than or equal to `10,000` and with more than `8` schools per square kilometer:

2. From the **Selection** menu, click on **Select By Attributes**. For **Layer**, make sure that the `Municipalities_PrimarySecondarySchools` layer is selected. For **Method**, the default **Create a new selection** option should be selected.

3. Click on the **Load** button, and select the **Suitable_Municipalities. exp** expression file from the `<drive>:\LearningArcGIS\Chapter7\ Geoprocessing` folder. Click on **OK**.

4. In **Table Of Contents**, right-click on the **Municipalities_ PrimarySecondarySchools** layer and select **Create Layer From Selected Feature** in **Selection**. This layer resides only within the `Site_selection` map document (.mxd), and its data source is the `Municipalities_ PrimarySecondarySchools` feature class.

5. Click on the new map layer named `Municipalities_ PrimarySecondarySchools selection` and drag it below the **Hospitals** layer. Click on its text again to edit it and change its name to `Suitable_ Municipalities`.

6. Let's adjust the transparency of **Suitable_Municipalities** to see the layers underneath it. Double-click on the layer to open **Layer Properties**. Click on the **Display** tab, and type `60` in the box next to **Transparent**. Click on **OK**.

7. Turn off the `Municipalities_PrimarySchools` and `Municipalities_ PrimarySecondarySchools` layers.

8.  Right-click on the `Suitable_Municipalities` layer and select **Open Attribute Table**. The layer table stores only three municipalities: `Saint-Josse-ten-Noode`, `Etterbeek`, and `Koekelberg`.

Next, we will select those primary and secondary schools within the three municipalities that are candidates for the new pediatric center. Here, we have two ways to solve the problem: Use **Select By Location** to select the schools and create two temporary layers in the map document, or use the **Clip** tool to create two different feature classes in the geodatabase, which store the primary and secondary schools.

Supposing that we don't want to create and store too many intermediary datasets in the `World.gdb` file geodatabase, we will choose the first option:

9.  From the **Selection** menu, click on **Select By Location**. For **Selection method**, make sure that the **select features from** option is selected. For **Target layer (s)**, select the **Primary_schools** and **Secondary_schools** layers. For **Source layer**, choose the **Suitable Municipalities** layer. For **Spatial selection methods**, select the **intersect the source layer feature** option. Click on **OK**.

10. In **Table Of Contents**, right-click on the **Primary_schools** layer and select **Create Layer From Selected Feature** from **Selection**. Change the name of the **Primary_schools selection** new layer to `Primary`. To change the symbol for the new layer, open **Layer Properties**. In the **Symbology** panel, click on the **Import** button at the top-right corner of the **Layer Properties** window, and for **Layer** drop-down list, select the **Primary_Schools** layer. Click on **OK** twice to close the dialog windows.

11. Repeat the previous step for the `Secondary_schools` layer. Change the name of the new layer to `Secondary`. Also, change its symbol.

12. Turn off the `Primary_schools` and `Secondary_schools` layers, as shown in the following screenshot:

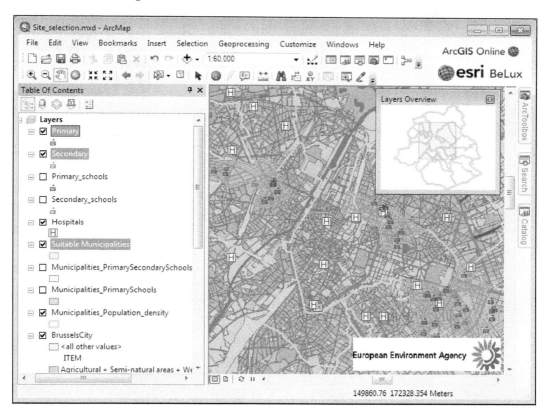

13. Next, we will create a 200 meters buffer around each school. From the **Geoprocessing** menu, click on the **Buffer** tool. In the **Buffer** dialog window, set the parameters, as shown in the following screenshot:

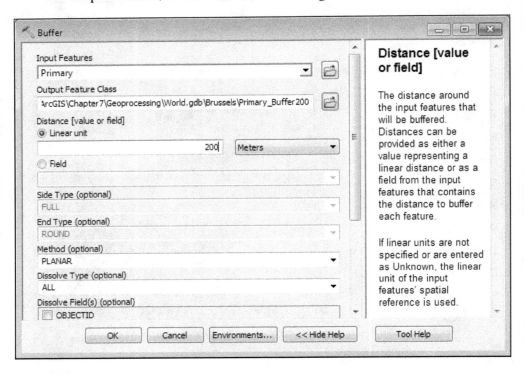

14. For **Input Features**, choose the **Primary** layer. For **Output Feature Class**, change the default name of the output feature class to `Primary_Buffer200`. Under **Distance**, type `200` for **Linear unit**, and select **Meters**.

15. For **Dissolve Type (optional)**, select **ALL** to dissolve all individual buffer polygons into a single one. Accept the default values for all other parameters.

16. Click on **OK** to run the tool. The `Primary_Buffer200` layer is added to the map.

17. From the **Geoprocessing** menu, click again on the **Buffer** tool. In the **Buffer** dialog window, set the following parameters:

    ○ **Input Features**: Set this to `Secondary`

    ○ **Output Feature Class**: Set this to `World.gdb\Brussels\Secondary_Buffer200`

    ○ **Linear unit**: Set this to `200 Meters`

    ○ **Dissolve Type (optional)**: Set this to `ALL`

18. Accept the default values for all other parameters. Click on **OK** to run the tool. The **Secondary_Buffer200** layer is added to the map.

19. Next, we will overlay the school buffer features to combine them into one layer. From the **Geoprocessing** menu, choose the **Union** tool and set the following parameters:

    ○ **Input Features**: Set this to add the `Primary_Buffer200` and `Secondary_Buffer200` layers

    ○ **Output Feature Class**: Set this to `World.gdb\Brussels\School_Buffer200`

    ○ **JoinAttributes (optional)**: Set this to `ALL`

    ○ **Gaps Allowed (optional)**: Make sure that this is selected

20. Click on **OK** to run the tool. The **School_Buffer200** layer is added on the map. Right-click on it and choose **Zoom To Layer**.

21. Turn off the `Primary_Buffer200` and `Secondary_Buffer200` layers.

22. Let's adjust the symbol and transparency of the **School_Buffer200** layer. Double-click on the layer name to open its **Layer Properties** window. Click on the **Display** tab and set **Transparent** to `40`. Click on the **Symbology** tab and then click on the symbol. In the **Symbol Selector** window, change the **Fill Color** to **Yucca Yellow**. Click on **OK**.

23. Next, we will create a 600 meters buffer around each hospital. From the **Geoprocessing** menu, choose the **Buffer** tool and set the following parameters:

    ○ **Input Features**: Set this to `Hospitals`

    ○ **Output Feature Class**: Set this to `World.gdb\Brussels\Hospitals_Buffer600`

    ○ **Linear unit**: Set this to `600 Meters`

    ○ **Dissolve Type (optional)**: Set this to `ALL`

24. Accept the default values for all other parameters. Click on **OK** to run the tool. The `Hospitals_Buffer600` layer is added to the map.

25. To identify the areas that meet the criteria for the pediatric center site, we will overlay the school buffer features with the `Hospitals_Buffer600` layer. From the **Geoprocessing** menu, choose **ArcToolbox**. Dock the **ArcToolbox** on the right-hand side of the ArcMap window. In **ArcToolbox**, expand the **Analysis Tools** *toolbox*. Next, expand the **Overlay** *toolset*, and double-click on the **Erase** *system tool* to open its dialog window. Set the following parameters:

    ○ **Input Features**: Set this to `Schools_Buffer200`

    ○ **Erase Features**: Set this to `Hospitals_Buffer600`

- ○ **Output Feature Class**: Set this to `World.gdb\Brussels\Critical_Zones`

26. Click on **OK** to run the tool. The `Critical_Zones` layer is added to the map. These are the critical areas that need to be covered by the services of a pediatric center.

27. Turn off the `Schools_Buffer200` and `Hospitals_Buffer600` layers.

28. Right-click on the **Critical_Zones** layer and choose **Zoom To Layer**. Change its default symbol color to **Mars Red** and the transparency to `50%`, as shown in the following screenshot:

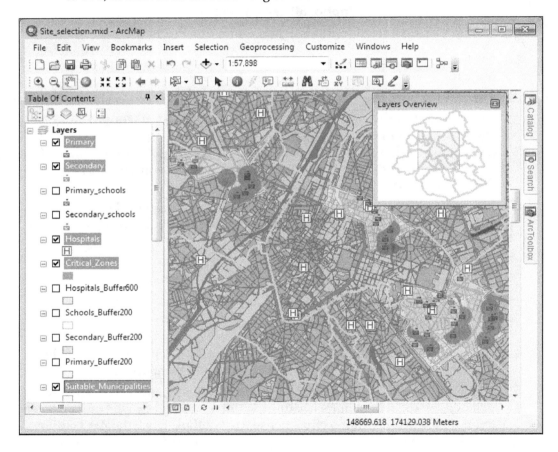

29. Save the `Critical_Zones` layer as **Layer File** (right-click on layer and choose **Save As Layer File**).

In the last part of this exercise, we will identify the parcels without current use from the `BrusselsCity` layer that are available in each of the critical areas. To select the parcels without current use, we have three ways to solve the problem: use **Select By Attributes** to select the parcels without current use, use **Query Builder** to display only those parcels without current use on the map, and use the **Select** geoprocessing tool from **ArcToolbox** to create a new feature class that stores the parcels without current use. To explore more geoprocessing tools, we will choose the last option:

30. In ArcToolbox, expand **ArcToolbox | Analysis Tools | Extract** and double-click on the **Select** tool to open its dialog window. Set the following parameters:

    ° **Input Features**: Set this to `BrusselsCity`

    ° **Output Feature Class**: Set this to `World.gdb\Brussels\Parcels`

    ° **Expression (optional)**: Set this to `ITEM = 'Land without current use'`

31. Click on **OK** to run the tool. The `Parcels` layer is added to the map.

32. From the **Selection** menu, click on **Select By Location** to open the dialog window. Set the following parameters:

    ° **Selection method**: Set this to `select features from`

    ° **Target layer(s)**: Set this to `Parcels`

    ° **Source layer**: Set this to `Critical_Zones`

    ° **Spatial selection method**: Set this to `intersect the source layer feature`

33. Click on **OK**. Open the attribute table of the `Parcels` layer and click on the **Show selected records** button. There is only one parcel that meets all specified criteria to build a pediatric center, as you can see in the following screenshot:

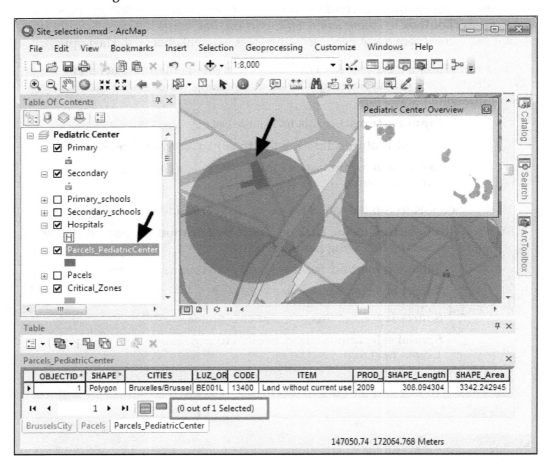

34. Right-click on the **Parcel** layer and select **Export Data** from **Data** to save the selected parcel in a new feature class named `Parcels_PediatricCenter`. Click **Yes** to add the new feature class as a layer on the map.

35. From the **View** menu, select **Data Frame Properties**. Click on the **General** tab, and change the **Name** of the feature class to `Pediatric Center`. Click on **OK**. Note that the **Overview** name has changed. To view `Critical_Zones` as a background in the `Pediatric Center Overview` window, right-click on the top blue section of the window, and choose **Properties**. For **Reference layer**, select **Critical_Zones**. Click on **OK**.

36. When finished, save your changes in a new map document named `MySite_Selection.mxd` to `<drive>:\LearningArcGIS\Chapter7\Geoprocessing`.

37. Leave the map document open to learn how to share your analysis results.

You can find the results at `<drive>:\LearningArcGIS\Chapter7\Results\Geoprocessing`.

# Sharing the analysis results

You can share the results of your analysis by exporting the maps as the following:

- A PDF or image file for non-ArcGIS users
- A map package (`.mpk`) uploaded to the ArcGIS Online account or saved on the local disk for ArcGIS users
- Services on the ArcGIS Server

A map package file includes the map document (`.mxd`), a copy of the file geodatabase, and any other additional files. Data stored in a package compressed file (`.mpk`) can be shared with other users on the local network, ArcGIS Online, or by e-mail. The map package can be downloaded and unpacked on the local disk.

Follow these steps to prepare the results of the site selection analysis to be shared:

1. If necessary, start the ArcMap application and open your map document `MySite_Selection.mxd` from `..\Chapter7\Geoprocessing`.

2. Before you start sharing your data, you should document your map document. From the **File** menu, select **Map Document Properties**. For **Title**, type `Pediatric center site`. For **Summary**, type `The aim of this site selection analysis is to find the most suitable parcels for a new Pediatric Center in the city of Brussels`.

3. For **Description**, type `The map documents contents layers, representing the parcels, municipalities, hospitals, and primary and secondary schools in the city of Brussels`.

4. For the **Author** and **Credits** fields, type `European Environment Agency` and `Esri_Belux_Content`. For **Tags**, type `population, schools, hospitals`. Click on **Apply** and then click on **OK**.

 If you want to share your analysis results on ArcGIS Online, in the **File** menu, select **Sign in**. Use your username and password to sign in to your ArcGIS Online account. In the **File** menu, navigate to **Share As | Map Package**. Check **Upload package to my ArcGIS Online account** and give a name to your map package.

5. Next, we will save the analysis results as a map package on the local disk. From the **File** menu, select **Map Package** in **Share As**. In the **Map Package** panel, for **Save package to file**, navigate to `..\Chapter7\Geoprocessing`, and save the file as `MySite_Selection.mpk`. Click on **Save**.

6. Click on the second panel named **Item Description**. The required fields are already filled out with information from **Map Document Properties**.

7. Use the **Additional Files** panel if you want to add more files to your map package.

8. When finished, click on the **Analyze** button on the top-right corner of the **Map Package** window to locate all possible errors in your map document, such as missing descriptive information, missing spatial coordinate system, or layers with inaccessible data sources.

9. As you can see in the **Prepare** window, there are no reported errors. Your map package is ready to be shared.

10. Click on the **Share** button at the top-right corner of the **Map Package** window. Click on **Yes** to save the changes to your map document.

11. When finished, click on **OK**. Close ArcMap.

12. Open the Windows Explorer application and navigate to `<drive>:\LearningArcGIS\Chapter7\Geoprocessing`. Expand the **Geoprocessing** folder and double-click on **MySite_Selection.mpk**.

13. The `MySite_Selection.mxd` map document should open. Open the **Map Document Properties** window, and check the location of the unpacked data.

14. In the **Catalog** window, connect to the **MySite_Selection** folder, as shown in the following screenshot:

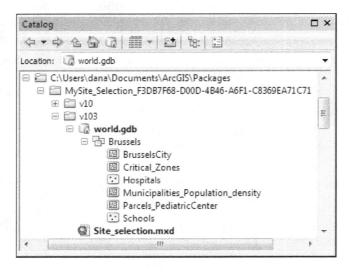

15. Explore the data. When finished, close the ArcMap application.

You can find the results at <drive>:\LearningArcGIS\Chapter7\Results\
Geoprocessing.

# Working with ModelBuilder

The **ModelBuilder** built-in application is a graphical environment that allows you to
visualize, create, and run a geoprocessing workflow or geoprocessing model. In the
context of geoprocessing, a model contains a GIS process or a sequence of connected
GIS processes, as shown in the following screenshot:

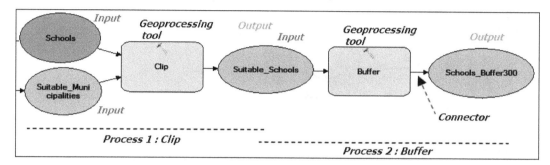

A GIS process consists of the following:

- A geoprocessing tool, such as the **Buffer** tool or the **Clip** tool (the orange
  elements in the previous screenshot)

- One or more input datasets (the blue elements in the previous screenshot)

- An output dataset (the green element in the previous screenshot)

As you may have noticed in the previous screenshot, in a geoprocessing model,
an output dataset of one process can become an input dataset for the next related
process. An output dataset that became the input dataset for another process is
called **intermediate data**.

# Creating a model

In a geoprocessing model, you can use any of the **ArcToolbox** tools or custom tools that you created using ModelBuilder or Python. When you create a geoprocessing model in ModelBuilder, you work with three basic types of elements:

- Data Variables (for example, input data, output data, or derived data) and Value Variables (for example, the buffer distance value)
- Tools (for example, system tools, such as the **Buffer** tool)
- Connectors (for example, the data connector and the environment setting)

In the next screenshot, you can see three different stages of a model in the ModelBuilder application:

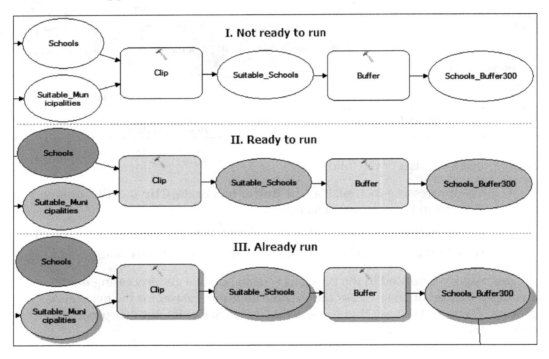

When the model elements are *white*, the processes are not ready to run. In this case, you should check whether the tool parameters are properly set and whether the input data contains errors. When the model elements are *colored*, the related processes are ready to run. When the tools (orange elements) and derived data (green elements) are *shadowed*, this means that the model processes have run successfully.

In the next exercise, we will perform the spatial analysis presented in the *Working with Geoprocessing tools* section again. To preview the proposed geoprocessing workflow, open the PDF file named `Pediatric Center in Brussels.pdf`, which is stored at `<drive>:\LearningArcGIS\Chapter7\ModelBuilder`. The workflow diagram displays the following tasks:

- Identifying the Brussels' municipalities with a population density of over `10,000` inhabitants and with more than `8` schools per square kilometer using the **Spatial Join** and **Select** tools

- Identifying the schools within the municipalities selected in the first step using the **Clip** tool

- Creating a `300` meters buffer around each school within the suitable municipalities and a `600` meters buffer around each hospital using the **Buffer** tool

- Identifying those critical areas that are not close to a hospital and need pediatric services using the **Erase** tool

- Identifying those parcels without current use and which intersect with the critical areas using the **Select** and **Select Layer By Location** tools

- Saving the suitable parcels to build a pediatric center in a new feature class called `Parcels_PediatricCenter` using the **Copy Features** tool

Follow these steps to start building a model called `Pediatric Center in Brussels` using the ModelBuilder built-in application:

1.  Start the ArcCatalog application. In **Catalog Tree**, navigate to `<drive>:\LearningArcGIS\Chapter7\ModelBuilder` and expand it. Right-click on the **World.gdb** file geodatabase, and choose **Toolbox** in **New**.

> A model must be created in a custom toolbox in a Geodatabase or as a standalone `.tbx` file. A model cannot be saved to a system toolbox.

2.  Right-click on the newly created toolbox and select **Rename**. Change the toolbox's name to `MyToolbox`. Right-click on **MyToolbox** again and choose **Model** in **New** to open the ModelBuilder window.

Before we start adding any processes to the empty model, let's set the main parameters of **Environments**:

3.  From the **Model** menu, select **Model Properties**. The **General** tab should be selected. For **Name**, type `PCenter`. For **Label**, type `Pediatric Center in Brussels`.

4. It's good practice to document your model. For **Descriptions**, type the following text: The aim of this site selection analysis is to find the most suitable parcels for a new Pediatric Center in the city of Brussels.

5. Select the **Store relative path names (instead of absolute paths)** option. The relative path is relative to the current folder, Chapter7\ModelBuilder, which stores the World.gdb file geodatabase and the PCenter model. If you moved or copied your ModelBuilder folder anywhere on a computer or any data storage device and you keep the same folder and geodatabase structure, the PCenter model will find the referenced datasets and will use them.

6. Click on the **Environments** tab. First, select the **Output Coordinates** and **Workspace** categories. Click on the **Values** button. Expand the **Workspace** category, and for **Current Workspace**, select <drive>:\LearningArcGIS\ Chapter7\ModelBuilder\World.gdb\Brussels. For **Scratch Workspace**, select <drive>:\LearningArcGIS\Chapter7\ModelBuilder\World.gdb\ Brussels again.

7. Expand **Output Coordinates**, and make sure that the **Same as Input** option is selected. Click on **Apply** and then click on **OK**.

8. Next, from the **Insert** menu, select **Add Data or Tool**. From the **Look in** drop-down list, select **Toolboxes**. Double-click on **System Toolboxes**, navigate to **Analysis Tools | Overlay**, and select **Spatial Join**. Click on **Add**.

9. In the model, double-click on the **Spatial Join** tool and set the following parameters:

   ° **Target Features**: Set this to Chapter7\ModelBuilder\World.gdb\ Brussels\Municipalities_Population_density

   ° **Join Feature**: Set this to Schools

   ° **Output Feature Class**: Set this to Population_Schools_Density

   ° **Join Operation (optional)**: Set this to JOIN_ONE_TO_ONE

   ° **Keep All Target Features (optional)**: Keep this selected

   ° **Match Options (optional)**: Set this to INTERSECT

10. Accept the default values for all other parameters. Click on **OK**. The Municipalities_Population_density and Schools elements were added to the process. Save the changes to the model using the **Save** button from the **Standard** toolbar.

Next, we will add the second process shown in the workflow diagram in the PDF file named `Pediatric Center in Brussels.pdf`:

11. From the **Insert** menu select **Add Data or Tool**. From the **Look in** drop-down list, select **Toolboxes**. Navigate to **System Toolboxes | Data Management Tools | Fields** and select **Add Field**. Click on **Add**.

12. In the model, double-click on the **Add Field** tool and set the following parameters:
    ○ **Input Table**: From the drop-down list, select the output data element named `Population_Schools_Density`
    ○ **Field Name**: Set this to `Schools_Density`
    ○ **Field Type**: Set this to `SHORT`

13. Accept the default values for all other parameters. Click on **OK**. Save the changes to the model.

14. Next, we will calculate the values for the `Schools_Density` field. On the **Standard** toolbar, click on the **Search** button. In the **Search** window, search for `calculate field` to locate the **Calculate Field (Data Management)** tool. Drag it in the model.

15. Double-click on the **Calculate Field** tool and set the following parameters:
    ○ **Input Table**: From the drop-down list select the output data element named `Population_Schools_Density (2)`
    ○ **Field Name**: Set this to `Schools_Density`
    ○ **Expression**: Set this to `[Join_Count]/[Area_km2]`

16. Accept the default values for all other parameters. Click on **OK**. Save the changes to the model.

Next, we will identify the Brussels' municipalities with a population density of over `10,000` inhabitants and with more than `8` schools per square kilometer:

17. In the **Search** window, search for the **Select (Analysis)** tool. Drag it into the model. Double-click on the **Select** tool and set the following parameters:
    ○ **Input Features**: Select the output data element named `Population_Schools_Density (3)`
    ○ **Output Feature Class**: Set this to `Suitable_Municipalities`
    ○ **Expression**: Set this to `Density_km >10000 AND Schools_Density >8`

18. Click on **OK**. Save the changes to the model.

Next, we will identify the schools within the suitable municipalities:

19. In the **Catalog** window, expand **System Toolboxes | Analysis Tools | Extract**. Select the **Clip** tool and drag it into the model.

20. Double-click on the **Clip** tool and set the following parameters:

    ○ **Input Features**: From the drop-down list, select the input data element named `Schools`

    ○ **Clip Features**: Select the output data element named `Suitable_Municipalities`

    ○ **Output Feature Class**: Set this to `Suitable_Schools`

21. Click on **OK**. Save the changes to the model.

22. Use the **AutoLayout** button from the **Standard** toolbar to rearrange the model elements. To see the entire model, you should slightly enlarge the ModelBuilder window and use the **Full Extent** button.

Next, we will create a 300 meters buffer around each school within the suitable municipalities and a 600 meters buffer around each hospital using the **Buffer** tool:

23. Navigate to **System Toolboxes | Analysis Tools toolbox | Proximity** and drag the **Buffer** tool next to the **Suitable_Schools** output data element. Add one more **Buffer** tool below the first one, as shown in the following screenshot:

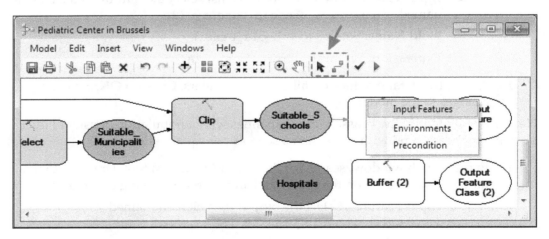

24. In the ModelBuilder window, click on the **Connect** tool. Click on the **Suitable_Schools** output data element, and then click on the **Buffer** tool. Click on **Input Features** in the displayed context menu.

25. Click on the **Select** tool, and double-click on the **Buffer** tool to set its parameters, as follows:

    ° **Input Features**: `Suitable_Schools` is already selected

    ° **Output Feature Class**: Set this to `Schools_Buffer300`

    ° **Distance**: Set this as **Linear unit** = `300` **meters**

    ° **Dissolve Type (optional)**: Set this to `ALL`

26. Accept the default values for all other parameters. Click on **OK**.

27. Double-click on the **Buffer (2)** tool and set its parameters, as follows:

    ° **Input Features**: This is `Chapter7\ModelBuilder\World.gdb\` `Brussels\Hospitals`

    ° **Output Feature Class**: This is `Hospitals_Buffer600`

    ° **Distance**: This is **Linear unit** = `600` **meters**

    ° **Dissolve Type (optional)**: This is `ALL`

28. Accept the default values for all other parameters. Click on **OK**. Save the changes to the model.

Next, we should identify those critical areas which are not close to a hospital and need pediatric services using the **Erase** tool:

29. Navigate to **System Toolboxes | Analysis Tools toolbox | Overlay** and drag the **Erase** tool next to the `Hospitals_Buffer600` output data element. Double-click on the **Erase** tool and set the following parameters:

    ° **Input Features**: From the drop-down list, select the output data element named `Schools_Buffer300`

    ° **Erase Features**: This is `Hospitals_Buffer600`

    ° **Output Feature Class**: This is `Critical_Zones`

    ° **Dissolve Type (optional)**: This is `ALL`

30. Click on **OK**. Save the changes to the model.

Next, we will identify those parcels without current use and which intersect with the critical areas, as shown in the following screenshot:

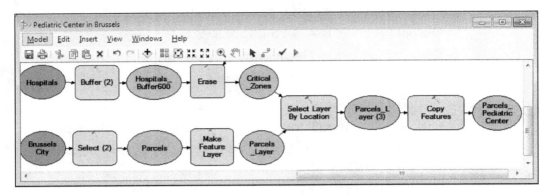

31. Navigate to **System Toolboxes | Analysis Tools toolbox | Extract** and drag the **Select** tool into the model. Double-click on the **Select** tool and set the following parameters:

    ◦ **Input Features**: This is `Chapter7\ModelBuilder\World.gdb\Brussels\BrusselsCity`

    ◦ **Output Feature Class**: This is `Parcels`

    ◦ **Expression (optional)**: This is `ITEM='Land without current use'`

32. Click on **OK**. Navigate to **System Toolboxes | Data Management Tools toolbox | Layers and Table Views** and drag the **Make Feature Layer** tool into the model.

33. With the **Connect** tool selected, click on the **Parcels** output data element, and then click on the **Make Feature Layer** tool. Click on **Input Features** in the displayed context menu.

34. Change the default name of the output data element to `Parcels_Layer`. Save the changes to the model.

35. Navigate to **System Toolboxes | Data Management Tools toolbox | Layers and Table Views** and drag the **Select Layer By Location** tool into the model.

36. Double-click on the tool and set the following parameters:

    ◦ **Input Feature Layer**: From the drop-down list, select `Parcel_Layer`

    ◦ **Relationship (optional)**: Set this to `INTERSECT`

    ◦ **Selecting Features (optional)**: Set this to `Critical_Zones`

37. Accept the default values for all other parameters. Click on **OK**.

In the last part of this exercise, we will save the suitable parcels to build a pediatric center in a new feature class called `Parcels_PediatricCenter`:

38. In the **Search** window, search for the **Copy Features (Data Management)** tool. Drag it into the model. Click on the **Connect** tool from the **Standard** toolbar. Click on the **Parcels_Layer (2)** output data element and then click on the **Copy Features** tool. Choose **Input Features**.

39. Rename the output data element to `Parcels_PediatricCenter`.

40. You should validate the entire model before you run it. From the **Model** menu, choose **Validate Entire Model**. The model is ready to run.

41. All green output data elements are considered intermediate data, except the last output data element named `Parcels_PediatricCenter`, which is the result of the `PCenter` geoprocessing workflow. Right-click on several of the output elements and note that **Intermediate** is selected.

 If you run the model in the ArcMap application, right-click on the last output element and select **Add To Display**. The `Parcels_PediatricCenter` final result will be automatically added to the map display.

42. Click on **Auto Layout** and **Full Extent** to see the whole geoprocessing workflow.

43. Save the changes to the model. Close **ArcCatalog**.

You can find the `PCenter` model at `<drive>:\LearningArcGIS\Chapter7\Results\ModelBuilder\World.gdb\Site_Selection_Analysis`.

# Using a model

In this subsection, we will validate and run the `PCenter` model that was created in the previous section. With a model created in ModelBuilder, you can automate a geoprocessing workflow, run it as one tool whenever necessary, and furthermore, share it with other users. It is good practice to validate the model before running or sharing it. The validation verifies and locates the errors in the model, such as incorrect parameters, improper values for parameters, inaccessible data sources, and missing spatial coordinate system for the data source.

Follow these steps to run the `PCenter` model created in the previous exercise:

1. Start the ArcCatalog application. In **Catalog Tree**, navigate to `<drive>:\LearningArcGIS\Chapter7\ModelBuilder\World.gdb`, and expand `MyToolbox`.

2. Right-click on the **Pediatric Center in Brussels** model, and choose **Open**. In the **Pediatric Center in Brussels** dialog window, you will see the following message: **This tool has no parameters.** As no parameters were defined in the PCenter model, it is normal to see this warning message. Click on **OK** to run the tool.

3. When finished, click on **Cancel**.

4. The final result is added to the Brussels feature dataset. To see the new feature classes, right-click on the **Brussels** feature dataset and select **Refresh**.

5. Select the **Parcels_PediatricCenter** feature class and click on the **Preview** tab to view the data.

 The intermediate output feature classes were automatically deleted in the Brussels feature dataset because you ran the model from its tool dialog window.

6. Remove the Parcels_PediatricCenter feature class from the Brussels feature dataset to run the model again using the ModelBuilder application. Right-click on the **Pediatric Center in Brussels** tool and choose **Edit** to open the ModelBuilder window.

7. Move the **Pediatric Center in Brussels** window so that you can see the model and the file geodatabase content in **Catalog Tree** at the same time.

8. From the **Model** menu, click **Run Entire Model**. Take note of how the tools are colored in red while they are running one after the other.

9. When finished, note that the tools (orange) and derived data elements (green) have shadows. This means that the model is in the third stage: *The tools have already been run.*

 Also, take note of the intermediate feature classes from the Brussels feature dataset in **Catalog Tree**. The intermediate output feature classes were added in the Brussels feature dataset because you ran the model from ModelBuilder.

10. If you want to remove the intermediate feature classes, go to the **Model** menu and choose **Delete Intermediate Data**. In the ModelBuilder window, note that the model returned to the second stage: *Ready to run.*

11. Another way to remove the shadows (return to the second stage of model) is to validate the model using the **Validate Entire Model** tool from the **Standard** toolbar.

12. If you want to run only a part of the model, use the **Select** tool from the **Standard** toolbar to select one or two connected processes from the model. Then, right-click on the selected processes and choose **Run**.

13. To delete the output feature classes from file geodatabases, use the **Delete Intermediate Data** option again. Save and close the model.

In the last part of this exercise, we will run the model in the ArcMap application:

14. Open ArcMap. Add a new empty map document.

15. Open the **Catalog** window and navigate to your **Pediatric Center in Brussels** tool.

    Document your model using **Item Description**. Right-click on the **Pediatric Center in Brussels** tool and choose **Item Descriptions**. Document your tool using the knowledge you have gained in *Chapter 3, Creating a Geodatabase and Interpreting Metadata*.

    You can annotate the tools and data elements from your model. To add graphic labels to your model, use **Create Label** from **Insert Menu**.

16. Right-click on the tool and select **Edit** to open the ModelBuilder window.

17. On the **Standard** toolbar, click on the **Run** button.

18. When finished, the `Parcels_PediatricCenter` feature class is automatically added to the map.

19. Save and close the model. Save your map and close the ArcMap application.

# Summary

In this chapter, you saw that geoprocessing is an integral part of any spatial analysis and consists of creating new data from existing data. You chose and combined the appropriate geoprocessing tools to perform a site selection analysis. You saw that **ArcToolbox** contains toolboxes that are containers for tools and specific geoprocessing functionality toolsets. The toolsets are logical containers for tools. A tool performs a single geoprocessing operation.

In the second part of this chapter, you automated the site selection analysis by creating and running a model with the ModelBuilder application. You can create a model in a custom toolbox and store it in a file geodatabase, in a folder on the local disk, or in **ArcToolbox**. The main advantages of using a model are that it can be run as many times as you need, it offers a visual overview of the analysis workflow, and it can be shared with other ArcGIS users.

In the next chapter, you will use the Spatial Analyst and 3D Analyst extensions to create and analyze raster data.

# 8

# Using Spatial Analyst and 3D Analyst

In this chapter, we will explore the analytic tools added by two software modules named Spatial Analyst and 3D Analyst. These two modules or extensions are optional, and can be used with all the Desktop license levels of ArcGIS: **Basic**, **Standard**, and **Advanced**. We will show you how to use the Spatial Analyst tools to perform a site selection, and a least-cost path analysis using raster data. We will also use the 3D Analyst extension to create, symbolize, and analyze three-dimensional data, such as 3D Features and TINs.

By the end of this chapter, you will have learned the following:

- Setting the data analysis environment
- Converting data
- Reclassifying raster data
- Performing a least-cost path analysis
- Creating a TIN surface
- Creating 3D features from 2D features
- Calculating surface area and volume

# Using Spatial Analyst

This section introduces the main types of analyses that you can perform on the cell-based data or raster data model using the ArcGIS Spatial Analyst extension. The Esri GIS Dictionary defines the raster as "an array of equally-sized cells arranged in rows and columns, and composed of single or multiple bands" (*A to Z GIS: An Illustrated Dictionary of Geographic Information Systems, Tasha Wade and Shelly Sommer, Esri Press*). The raster data represents a feature through a squared cell (or pixel), or a group of squared cells (pixels) with the same attribute value. The size of a cell defines the spatial resolution of a raster. A raster with small cells has a higher resolution, while a raster with large cells has a lower resolution. A raster can be stored as follows:

- It can be stored as a raster dataset inside a geodatabase.
- Outside a geodatabase, it can be stored as a raster dataset on disk. It may be stored in many formats, such as ESRI Grid, TIFF, JPEG, MrSID (Multiresolution Seamless Image Database), or IMG (ERDAS Image)

Throughout this chapter, we will work with discrete and continuous thematic raster datasets stored inside the file geodatabase.

A discrete thematic raster stores categorical or discontinuous data, such as land use, buildings, or hillshade raster. Discrete data can be *integer* (most commonly) or real number values. A discrete thematic raster has an associated **Value Attribute Table (VAT)** that contains the cell value in the Value field, and the number of cells for each discrete value in the Count field.

A continuous thematic raster stores continuous data or surface data, such as elevation, slope, or aspect (down-slope direction). Continuous data can be integer or *real number* (most commonly) values. In a continuous raster, each cell has a different value, though this is not necessarily unique. For example, it is possible for an elevation surface to have two or more cells with the same elevation value.

Also, a raster can store a NoData value to distinguish the cell with missing data from the cell with zero (0) value.

This section does not undertake to cover all the fundamental concepts of raster data, because the subject is very well documented in the ArcGIS Resource Center, and it is beyond the scope of this book. For more information about raster data, please refer to: http://resources.arcgis.com/en/help/.

Navigate to **Desktop (ArcMap): 10.4 | Manage Data | Data types | Raster and images**

# Performing suitability analysis

Suitability analysis (or site selection) is typically performed for analysis of change in land use, retail site selection, habitat, or real estate analysis, and so on.

In *Chapter 7, Working with Geoprocessing Tools and ModelBuilder,* you got acquainted with the main steps for performing spatial analysis, and we executed a site selection analysis using vector data. In this section, we will use the cell-based datasets (raster) and the ArcGIS Spatial Analyst tools to perform an analysis of change in land use.

Suppose the local community of a small village wants to increase the local orchard production by planting fruit trees, especially in those areas which are susceptible to landslides. The aim of the site selection analysis is to find the best places for the new fruit orchards. The suitable sites should meet the following conditions:

- The elevation should be over 225 meters
- The fruit orchards should have eastern and southern exposures — the direction of the terrain slope should be between 100 and 200 degrees
- The slope of the terrain should be between 10% and 40%

The coordinate reference system of the raster datasets is `Pulkovo 1942 Adj 58 Stereo 70`, with an azimuthal projection named `Double Stereographic` map projection and the `D_Pulkovo_1942_Adj_1958` datum.

The following workflow diagram shows the main tasks to be accomplished:

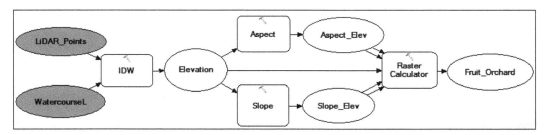

Follow these steps to perform a binary site selection using the ArcMap application and Spatial Analysis extension, as shown in the preceding workflow:

1.  Start the ArcMap application, and open the existing map document `Site_selection.mxd` from `<drive>:\LearningArcGIS\Chapter8\SpatialAnalyst`. The map document contains a layer that references a multipoint feature class that stores 3D data (*x*, *y*, and *z* values). It may take a minute or two to draw the `LiDAR_Points` layer. As you can see in the following screenshot, a multipoint feature class stores two or more points per row:

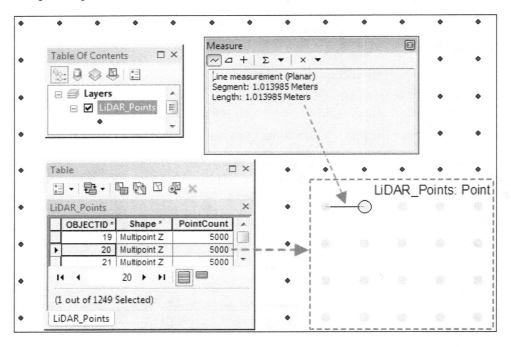

Since the `LiDAR_Points` layer, coming from the airborne laser scanning data (**LiDAR point cloud data**), may contain tens of thousands or millions of points, a multipoint feature class will reduce the storage space into a file geodatabase, and will improve the reading and writing of the point features. The `LiDAR_Points` layer stores regular grid points with a width of `1` meter.

2.  To make the extension available for the ArcMap application, select **Extensions** from the **Customize** menu, and check the **Spatial Analyst**. Click on **Close**. If you want to use the analysis tools of the Spatial Analyst extension in the ArcCatalog application, you should make the extension available using the same method that you used for the ArcMap application.

3. In the **Geoprocessing** menu, select **Environments**. Set the geoprocessing environment as follows:

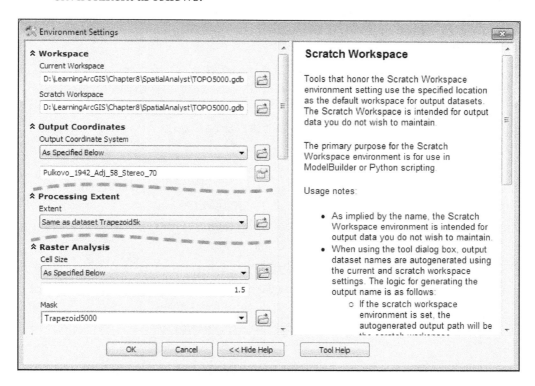

- ° In **Workspace** option, set both **Current Workspace** and **Scratch Workspace** as `<drive>:\LearningArcGIS\Chapter8\SpatialAnalyst\TOPO5000.gdb`

- ° **Output Coordinates**: Select the `LiDAR_Points` feature class from `..\Chapter8\SpatialAnalyst\TOPO5000.gdb\Relief`

- ° **Processing Extent**: Select the `Trapezoid5k` standalone feature class from `..\SpatialAnalyst\TOPO5000.gdb`

- ° **Raster Analysis**

- ° **Cell size: As Specified Below**: Select `1.5`

- ° **Mask**: Select `Trapezoid5k`

4. Click on **OK**. From the **Customize** menu, choose **Extensions**. Check the **Spatial Analyst** box, and click on **Close**. From the **Geoproccesing** menu, choose **ArcToolbox**, and dock it on the right side of the ArcMap window.

The most common data format for the storage of a **Digital Terrain Model (DTM)** is the raster format. In a raster, the cell value represents the height of the covered area. Next, we will use the interpolate raster surface using the **Inverse Distance Weighted (IDW)** approximation method, because the LiDAR_Points layer contains dense and evenly spaced points:

5. In the **ArcToolbox** window, expand the **Spatial Analyst Tools** toolbox. Expand the **Interpolation** toolset, and double-click on the **IDW** tool. Set the following parameters:

    ○ **Input point features**: Select the LiDAR_Points layer from the dropdown list

    ○ **Z values field**: Select Shape.Z

    ○ **Output raster**: Choose <drive>:\LearningArcGIS\Chapter8\ SpatialAnalyst\TOPO5000.gdb\Elevation

    ○ **Output cell size (optional)**: The default value is 1.5

    ○ **Power (optional)**: Enter 2

    ○ **Search radius (optional)**: Select Fixed

    ○ Under the **Search Radius Settings** set the parameters, **Number of Points** as 4 and **Maximum distance** as 2

    ○ **Input barrier polyline features (optional)**: Select ..\TOPO5000.gdb\ Hydrography\WatercourseL

6. Click on **OK** to run the tool. This tool may take 15 or more minutes to complete depending on your computer. When finished, the output raster is added to **Table Of Contents** as a new raster layer.

7. In **Table Of Contents**, turn off the LiDAR_Points layer.

8. Open **Layer Properties | Source** of the raster, and explore the following properties: **Format** (FGDBR = File Geodatabase Raster), **Cell Size** (1.5 meter), and **Spatial Reference** (Pulkovo_1942_Adj_58_Stereo_70). The Elevation raster is a *continuous thematic raster,* and the cells are arranged in 1553 rows and 1657 columns.

9. In the **Layer Properties** window, click on the **Symbology** tab. Select the **Show: Stretched** display method for the continuous raster cell values. For **Color Ramp**, choose **Surface**. Click on the **Display** tab, and set the **Transparency** parameter to 40%. Click on **OK**.

Next, we will create the necessary raster datasets for the site selection analysis. We will derive three rasters of hillshade, slope, and aspect from the `Elevation` surface:

 To preview the results of this exercise, please open the PDF files named `Hillshade`, `Slope`, and `Aspect`, stored at: `<drive>:\ LearningArcGIS\Chapter8\Results\PDFs`.

10. In the **ArcToolbox** window, expand the **Spatial Analyst Tools | Surface**, and double-click on the **Hillshade** tool. Set the following parameters:

    ° **Input raster**: select the `Elevation` layer from the dropdown list

    ° **Output raster**: Set as `.. \Chapter8\SpatialAnalyst\TOPO5000. gdb\Hillshade_Elev`

11. Accept the default values for the **Azimuth (optional)**, **Altitude (optional)**, and **Z factor (optional)** parameters. Click on **OK**.

12. In **Table Of Contents**, drag the **Hillshade_Elev** layer below the **Elevation** layer. The `Hillshade_Elev` layer produces a visually shaded effect that gives a 3D appearance to the `Elevation` layer in the ArcMap 2D environment.

13. The `Hillshade` raster is a *discrete thematic raster* that has an associated attribute table known as Value Attribute Table (VAT). Right-click on the **Hillshade** raster layer, and choose **Open Attribute Table**. The `Value` field stores the illumination values of the raster cells based on the position of the light source. The value `0` (black) means that the cells are not illuminated by the sun, and the value `254` (white) means that the cells are entirely illuminated. Keep the **Table** window open.

14. In the **ArcToolbox** window, expand the **Spatial Analyst Tools | Surface**, and double-click on the **Slope** tool. Set the following parameters:

    ° **Input raster**: Set as `Elevation`

    ° **Output raster**: Select `.. \Chapter8\SpatialAnalyst\TOPO5000. gdb\Slope_Elev`

    ° **Output measurement (optional)**: Select `PERCENT_RISE`

15. The slope will be expressed as a percentage (for example, 10%). Accept the default value for the **Z factor (optional)** parameter. Click on **OK**.

16. Symbolize the layer using the `Classified` method, as follows: **Show:** `Classified`. In the **Classification** window, select the value `5` from the drop-down list next to **Classes**. From the **Classification | Method** drop-down list, choose **Manual**.

17. In the **Break Values** section, change the first four values to 5, 10, 15, and 40. Leave the last break value, 110.13, which represents the maximum value, unchanged. Click on **OK**.

18. Select **Color Ramp**, and make sure the Slope (green to red) color ramp is selected. Click on **OK**.

19. In the **ArcToolbox** window, expand **Spatial Analyst Tools | Surface**, and double-click on the **Aspect** tool. Set the following parameters:

    ° **Input raster**: Set this as Elevation

    ° **Output raster**: Set this as .. \Chapter8\SpatialAnalyst\ TOPO5000.gdb\Aspect_Elev

20. Click on **OK**. The Aspect_Elev raster is added in **Table Of Contents** as a raster layer. Keep its default symbology.

Next, we will use the **Raster Calculator** tool to find the raster cells that meet all the criteria specified previously:

21. In the **ArcToolbox** window, expand **Spatial Analyst Tools | Map Algebra**, and double-click on the **Raster Calculator** tool. To determine the suitable raster cells, build the expression ("Elevation">225) & ("Aspect_Elev">= 100) & ("Aspect_Elev" <= 200) & ("Slope_Elev">= 10) & ("Slope_ Elev"<= 40), as shown in the following screenshot:

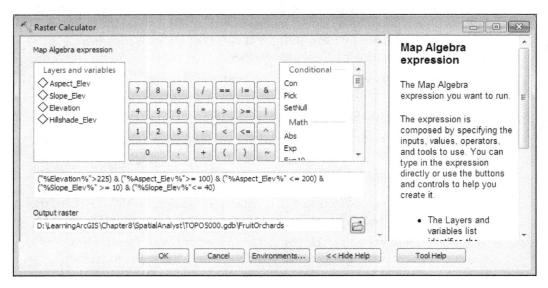

22. For **Output raster**, set `..\SpatialAnalyst\TOPO5000.gdb\Fruit_Orchards`. Click on **OK** to run the tool and create the `Fruit_Orchards` raster, as shown in the following screenshot:

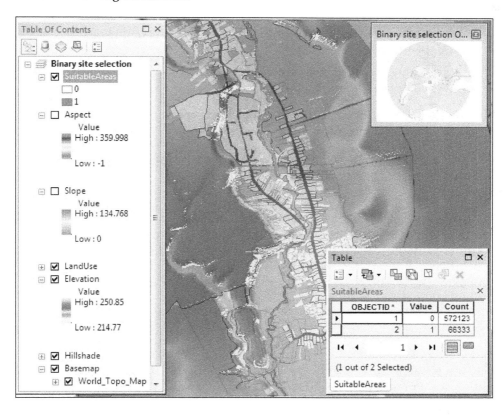

23. The `Fruit_Orchards` raster is a *discrete thematic raster*. Open its VAT by right-clicking on the raster layer, and choose **Open Attribute Table**. The `Value` field stores only two values: `0` or `1`. The value `0`, or logical false, means that the cells are not suitable. The value `1`, or logical true, means that the cells meet all the criteria, thus representing the suitable sites for a fruit orchard.

24. When finished, save your changes in a new map document named `MySite_Selection.mxd` in `<drive>:\LearningArcGIS\Chapter8\SpatialAnalyst`.

25. Close ArcMap.

You can find the results at: `<drive>:\LearningArcGIS\Chapter8\Results\SpatialAnalyst\Site_Selection.mxd`.

# Performing a least-cost path analysis

Two types of distance analysis can be made between two points, using the cell-based datasets (raster) as input. They are as follows:

- Determining the shortest distance using the Euclidean distance (straight-line distance)
- Determining the most cost-effective path using the cost-weighted distance and direction surfaces

In this subsection, we will refer only to the second type of distance analysis known as **weighted distance** or **least-cost path analysis**.

You could determine multiple paths between two points by taking into account (weighted) different factors that influence the *cost of travel* between a source and a destination.

Let's assume that we want to determine the trail that costs us the least amount of time and effort, from the center of a small village to two mountain cottages. The *factors* that could affect the travel across the area may be the slope of the terrain, the type of land use, and the snow depth. All these cost factors can be used to weigh the multiple paths that can be determined, and find the best path.

A *cost surface* represents one or more factors which are ranked using a common scale of weights.

For our scenarios, we should create a cost surface for each factor: cost of slope, cost of land use, and cost of snow depth. All cost surfaces must use a common weighting scale in order to combine them in a *total cost surface*.

For example, we can create a cost surface for the slope factor by ranking the cell values of the slope surface using a *1 to 5 weighting scale*. Since a steeper slope will cost effort and time, the highest slope values are assigned a value of 5, which means a higher travel cost. The lowest slope values are assigned a value of 1, which means a lower travel cost.

The total cost surface is used in the cost-weighted analysis to create the cost-weighted distance and direction surfaces. These two surfaces are used to determine the least-cost path from the center of the town (source) to a cottage (destination).

Follow these steps to perform a least-cost path analysis using the ArcMap application and Spatial Analysis extension, as shown in the following workflow:

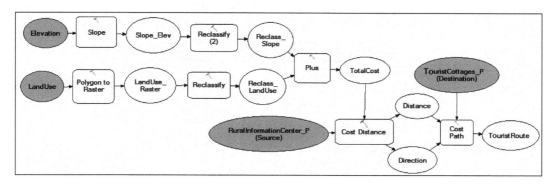

1. Start the ArcMap application, and open the existing map document Least-Cost Path.mxd from <drive>:\LearningArcGIS\Chapter8\ SpatialAnalyst. The map document contains two raster layers named Elevation and Hillshade as a basemap. Also, there are four vector layers, named WatercourseL, LandUse, RuralInformationCenter_P, and TuristCottages_P.

2. Since you will continue to work with the TOPO5000.gdb file geodatabase stored at <drive>:\LearningArcGIS\Chapter8\SpatialAnalyst, make sure the environment settings for this map document are the same as in the previous exercise.

3. In **Table Of Contents**, right-click on the **Elevation** layer, and choose **Properties**. Click on the **Source** tab, and click on the **Set Data Source** button. Navigate to <drive>:\LearningArcGIS\Chapter8\SpatialAnalyst\ TOPO5000.gdb, and select your Elevation raster created in the previous exercise. Click on **Add** and then on **OK**.

4. If necessary, open the **ArcToolbox** window, and dock it on the right-hand side of the ArcMap window.

First, we will reclassify the Slope_Elev raster created in the previous exercise:

5. In the **ArcToolbox** window, expand **Spatial Analyst Tools | Reclass**, and double-click on the **Reclassify** tool. Set the following parameters:

   ○ **Input raster**: Navigate to<drive>:\LearningArcGIS\Chapter8\ SpatialAnalyst\TOPO5000.gdb, and select Slope_Elev

   ○ **Reclass field**: Set as Value

- ° **Reclassification:** Click on the **Load** button, and choose `reclassslope` table from the `..\Chapter8\SpatialAnalyst` folder; the new values of the `Slope_Elev` raster have been loaded, as shown in the following screenshot:

| Old values | New values |
|---|---|
| 0 - 3 | 1 |
| 3 - 7 | 2 |
| 7 - 15 | 4 |
| 15 - 20 | 5 |
| 20 - 124.263977 | 5 |
| NoData | NoData |

- ° **Output raster:** Set as `..\Chapter8\SpatialAnalyst\TOPO5000.gdb\Reclass_Slope`

6. Click on **OK**. Inspect the new raster layer added on the map. When done, turn it off (uncheck it in **Table Of Contents**).

Next, we will convert the `LandUse` vector layer into a raster layer, and then reclassify its cell values:

7. In the **ArcToolbox** window, expand **Conversion Tools | To Raster**, and double-click on the **Polygon To Raster** tool. Set the following parameters:

- ° **Input Features:** Select the `LandUse` layer from the drop-down list
- ° **Value field:** Set as `LUCategory`
- ° **Output raster:** Set as `..\TOPO5000.gdb\LandUse_Raster`
- ° **Cell assignment type (optional):** Set as `MAXIMUM_COMBINED_AREA`

8. Accept the default values for all other parameters. Click on **OK** to run the tool. Once done, the output raster layer is added in **Table Of Contents**.

9. Turn off the `LandUse` layer. Right-click on the **LandUse_Raster** raster layer, and choose **Open Attribute Table**. Inspect the VAT fields and their values. Keep the **Table** window open.

10. In the **ArcToolbox** window, expand **Spatial Analyst Tools | Reclass**, and double-click on the **Reclassify** tool. Set the parameters as shown in the following screenshot:

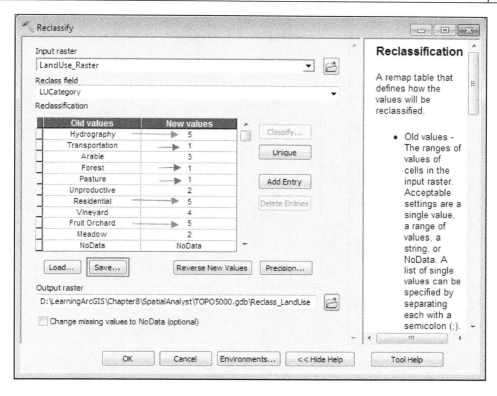

11. For **Input raster**, select **LandUse_Raster** raster layer. For **Reclass field,** select **LUCategory**. For **Reclassification**, click on the **Unique** button.

12. In the **New values** column, change the cell values as shown in the preceding screenshot. Use the **Save** button to save the new values in a remap table. For **Output raster**, type `Reclass_LandUse`.

13. Click on **OK**. Inspect the new raster layer added on the map. Open its Value Attribute Table, and inspect the field values. Keep the **Table** window open.

Next, we will create a total cost raster layer by combining the `Reclass_Slope` and `Reclass_LandUse` raster layers:

14. In the **ArcToolbox** window, expand **Spatial Analyst Tools | Math**, and double-click on the **Plus** tool. Set the following parameters:

    ° **Input raster or constant value 1**: Set this to `Reclass_Slope`

    ° **Input raster or constant value 2**: Set this to `Reclass_LandUse`

    ° **Output raster**: Set this as `..\TOPO5000.gdb\TotalCost`

15. Click on **OK**. When done, inspect the VAT of the `TotalCost` raster layer.

Next, we will derive two rasters, from the `TotalCost` raster layer, the cost depending on the distance from the source, and the cost depending on the direction of traveling, as shown in the following screenshot:

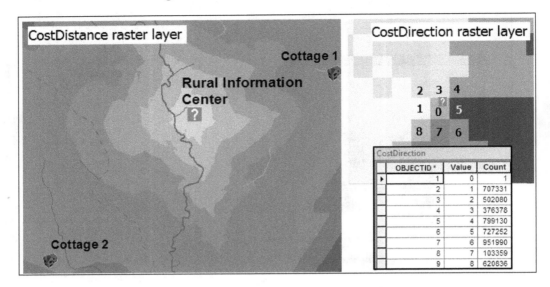

16. In the **ArcToolbox** window, expand **Spatial Analyst Tools | Distance**, and double-click on the **Cost Distance** tool. Set the following parameters:

    ○  **Input raster or feature source data**: Set this as `RuralInformationCenter_P`

    ○  **Input cost raster**: Set this as `TotalCost`

    ○  **Output distance raster**: Set as `..\TOPO5000.gdb\CostDistance`

    ○  **Maximum distance (optional)**: Select `<none>`

    ○  **Output backlink raster**: Set this as `..\TOPO5000.gdb\CostDirection`

17. Click on **OK**. The `CostDistance` layer represents how the cost increases as you drive away from the Rural Information Center, taking into account the distance measured from it and the total cost values from the `TotalCost` layer. The cell values represent the accumulative travel cost that can be a measure of the accessibility to the cottages based on the increasing slope of terrain and the distance from the source.

18. The `CostDirection` layer represents the direction to the least-cost path back to the nearest source (Rural Information Center), taking into account the total cost values from the `TotalCost` layer. Inspect the VAT of the `CostDirection` raster layer.

Based on these two cost rasters which have equal influence, we will find the least-cost paths between the Rural Information Center (source) and the tourist cottages (destination), as shown in the following screenshot:

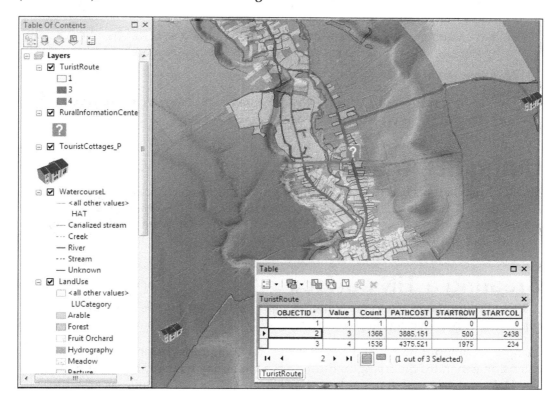

19. In the **ArcToolbox** window, expand **Spatial Analyst Tools | Distance**, and double-click on the **Cost Path** tool. Set the following parameters:

    ◦ **Input raster or feature destination data**: Input `TuristCottages_P`

    ◦ **Destination field (optional)**: Accept the default field

    ◦ **Input cost distance raster**: Set as `CostDistance`

    ◦ **Input cost backlink raster**: Input `CostDirection`

    ◦ **Output raster**: Set as `..\TOPO5000.gdb\TouristRoute`

    ◦ **Path type (optional)**: Set as `EACH_CELL`

20. Click on **OK**. When this is done, inspect the VAT of the `TouristRoute` raster layer. The `PATHCOST` field stores the smallest value of the accumulated cost for the two paths.

You can find the least-cost path from many sources to one destination.

21. Save your changes in a new map document named `MyLeastCost_Path.mxd` in `<drive>:\LearningArcGIS\Chapter8\SpatialAnalyst`.

22. Close ArcMap.

You can find the results at: `<drive>:\LearningArcGIS\Chapter8\Results\ SpatialAnalyst\LeastCost_Path`.

# Using 3D Analyst

The 3D Analyst tools can be used through four applications: ArcCatalog, ArcMap, ArcScene, and ArcGlobe. In this section, we will use the ArcScene application and 3D geoprocessing tools to display 2D features in a 3D perspective, and to create, convert, and analyze 3D data.

## Creating a TIN surface

A **TIN (Triangulated Irregular Network)** surface is a continuous elevation surface consisting of non-overlapping triangles.

Using a set of input points that contains the elevation information, the 3D Analyst extension generates the triangles through the **Delaunay triangulation** method. The Delaunay triangulation method creates triangles as equiangular as possible.

For more information about the TIN surface, please refer to the ArcGIS Resource Center: `http://resources.arcgis.com/en/help/`. Navigate to **Desktop (ArcMap): 10.4 | Manage Data | Data types | TIN**.

The input points or *mass points* can be point features (for example, LiDAR point data and peaks points) or vertices of the polyline or polygon features (contours as polylines). The elevation information is taken from an attribute field in the attribute table or from the Z-values stored in the 3D feature geometry.

Every triangle plane has one slope and aspect value. Within the triangle, you can find the elevation value for any $x$, $y$ coordinate location by interpolating the elevation value of the three nodes of the triangle, as shown in the following screenshot:

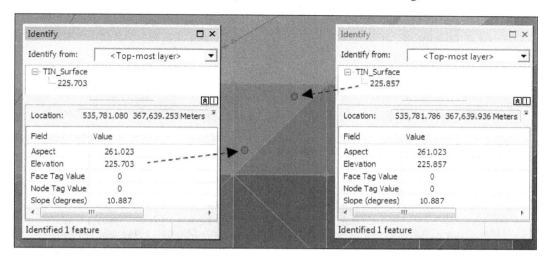

The following are the five surface feature types that can be added to a TIN to refine its surface:

- *Breaklines* enforces changes in slope: contours, rivers, roads, scarps, landslides, or building footprints as polyline or polygon features
- *Clip polygon* defines the area of interpolation where the elevation, slope, and aspect are calculated: study area boundaries as polygon features
- *Erase polygons* creates holes in the TIN surface: uncovered areas where the Z-value is missing, such as lakes, dense forest, or obscured surfaces
- *Replace polygons* creates flat areas on a TIN surface: building footprints or lakes with a constant height
- *Fill polygons* tags the triangle faces with supplementary attribute values coming from the various datasets: land use or vegetation as polygon features; thus, a TIN can be symbolized by other characteristics, such as land use types.

All these surface feature types can be added to a TIN as follows:

- *Hard* surface feature type: scarps, rivers as polylines, river or lake shorelines, roads, or building pads
- *Soft* surface feature type: study area, landslides, and contours

A TIN is stored directly in a folder structure, and cannot be stored in the geodatabase.

Follow these steps to create a TIN surface using the ArcGIS 3D Analyst extension in the ArcMap application:

1. Start the ArcMap application, and open the existing map document `CreatingTIN.mxd` from `<drive>:\LearningArcGIS\Chapter8\3DAnalyst`. The map document contains three layers representing the area of interest, rivers, and elevation points from LiDAR.

2. Open the **Catalog** window and connect to the folder, `Chapter8`.

3. From the **Customize** menu, choose **Extensions**. Check **3D Analyst** and click on **Close**.

4. From the **Geoprocessing** menu, choose **Environments**. For **Workspace | Current Workspace** and **Scratch Workspace**, select `<drive>:\LearningArcGIS\Chapter8\3DAnalyst\TOPO5000.gdb`. Click on **OK**.

5. From the **Geoprocessing** menu, choose **Geoprocessing Options**, and under **Background Processing**, make sure the **Enable** box is unchecked. In this way, all tools will be executed in foreground mode, which means that we should wait until the running tool returns us the output dataset (raster or vector data).

6. If necessary, check **Add results of Geoprocessing operations to the display box** under the **Display/Temporary Data** option. Click on **OK**.

7. If necessary, open the **ArcToolbox** window, and dock it on the right-hand side of the ArcMap window.

8. In the **ArcToolbox** window, expand **3D Analyst Tools | Data management | TIN**, and double-click on the **Create TIN** tool. Set the following parameters:

   ◦ **Output TIN**: Select `<drive>:\LearningArcGIS\Chapter8\3DAnalyst\TIN_Surface`

   ◦ **Coordinate System (optional)**: Choose `Pulkovo_1942_Adj_58_Stereo_70` from **Projected Coordinate Systems | National Grids | Europe**

   ◦ **Input Feature Class**: Click on the **Browse** button, and add the following three feature classes:

| Input Features | Height Field | SF Type | Tag Field |
|---|---|---|---|
| D:\LearningArcGIS\Chapter8\3DAnalyst\TOPO5000.gdb\Relief\LiDAR_Points | Shape.Z | Mass_Points | <None> |
| D:\LearningArcGIS\Chapter8\3DAnalyst\TOPO5000.gdb\Hydrography\WatercourseL | <None> | Hard_Line | <None> |
| D:\LearningArcGIS\Chapter8\3DAnalyst\TOPO5000.gdb\AOI | <None> | Soft_Clip | <None> |

9. Leave the last options unchecked, and then click on **OK** to run the tool. This tool may take a while to complete depending on your computer. When finished, **TIN_Surface** is added to **Table Of Contents**, as shown in the following screenshot:

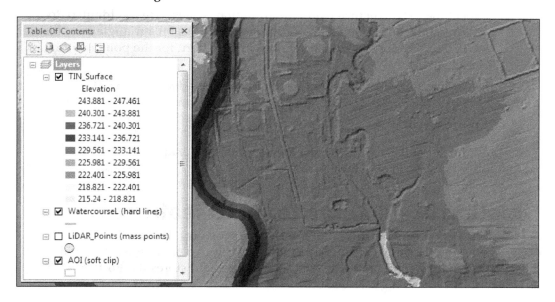

Next, we will resymbolize TIN_Surface, as shown in the following screenshot:

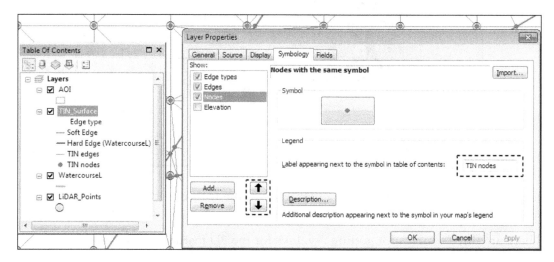

10. Double-click on **TIN_Surface** to open its **Layer Properties**. If necessary, click on the **Symbology** tab.

11. In the **Show** list, uncheck the **Elevation** box. Click on the **Add** button. In the **Add Renderer** window, select **Edges with the same symbol**. Click on **Add**. Next, select **Nodes** with the same symbol. Click on **Add**, and then click on **Dismiss**. Click on **OK**.

12. Inspect the TIN structure using different map scales. Use the **Identify** tool to determine the elevation value of different locations on a triangle face. Please notice that the slope and aspect values are constant for the points within a triangle plane.

13. Save your changes in a new map document named `MyCreatingTIN.mxd` at `<drive>:\LearningArcGIS\Chapter8\3DAnalyst`.

14. Close ArcMap.

You can find the results at: `<drive>:\LearningArcGIS\Chapter8\Results\3DAnalyst\CreatingTIN.mxd`.

# Creating 3D features from 2D features

3D features incorporate z-values into the feature geometry. When a 2D feature class is converted to a 3D feature class, and if the features are points, each of them gets a z-value along the *x*, *y* coordinates. The **Shape** field stores the **Point Z** value, as shown in the following screenshot:

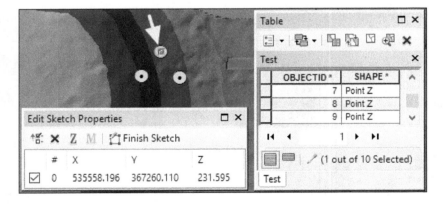

If the features are polylines or polygons, each of the feature vertices will get a z-value, and the `Shape` field will store the `Polyline Z` or `Polygon Z` value.

There are three sources for getting z-values for the 2D features, which are as follows:

- Using an attribute field that stores elevation information in the attribute table
- Using a constant z-value for all features

- Interpolating z-values from an underlying surface such as raster or TIN

 You can digitize the 3D feature into a 3D feature class in the ArcMap application using a reference surface (raster or TIN), and the 3D digitizing tools on the **3D Analyst** toolbar in an edit session.

In the next exercise, we will use the ArcScene application that is part of the 3D Analyst extension along with the ArcGlobe application. Follow these steps to convert 2D features to 3D using the 3D viewer named ArcScene:

1. Open the ArcScene application by navigating to **Start** | **All Programs** | **ArcGIS**. Click on **Browse** for more, and navigate to <drive>:\ LearningArcGIS\Chapter8\3D Analyst. Select the ArcScene document called 3D Features.sxd. The 3D scene contains three feature layers representing the rivers, roads, and a fish pond.

2. From the **Customize** menu, choose **Extensions**. Check **3D Analyst,** and click on **Close**.

3. From the **Geoprocessing** menu, choose **Environments**. For **Workspace** | **Current Workspace** and **Scratch Workspace**, select <drive>:\ LearningArcGIS\Chapter8\3DAnalyst\TOPO5000.gdb. Click on **OK**.

4. From the **Geoprocessing** menu, choose **Geoprocessing Options**, and make sure that the **Enable** box is unchecked under **Background Processing**.

5. If necessary, check **Add results of Geoprocessing operations to the display box** under **Display /Temporary Data**. Click on **OK**.

6. If necessary, open the **ArcToolbox** window, and dock it on the right side of the ArcScene window.

7. Click on the **Add Data** button on the **Standard** toolbar, and load the TIN_ Surface surface created in the previous exercise. On the **Tools** toolbar, use the **Navigate** tool to explore the TIN surface and the 2D feature layers.

 To set the base heights of 2D layers, navigate to **Layer Properties** | **Base Heights** | **Floating on a custom surface**.

8. In the **ArcToolbox** window, expand **3D Analyst Tools** | **Functional Surface**, and double-click on the **Interpolate Shape** tool. Set the following parameters:
   - **Input surface**: Select the TIN_Surface layer from the drop-down list
   - **Input Feature Class**: Set as Watercourse
   - **Output Feature Class**: Set as <drive>:\LearningArcGIS\ Chapter8\3DAnalyst\TOPO5000.gdb\Hydrography\Watercourse3D

      °    **Method (optional)**: Choose `LINEAR`

      °    Uncheck **Interpolate Vertices Only (optional)**

The **Interpolate Shape** tool will insert additional vertices along the river feature wherever it intersects the triangle edges, and will interpolate the z-value for every vertex, as shown in the following screenshot:

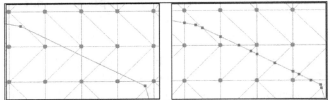

9.   Accept the default values for the rest of the parameters. Click on **OK**.

Let's explore the 2D and 3D feature vertices:

10. From the **Customize** menu, select **Toolbars,** and check the **3D Editor** toolbar. Click on **3D Editor**, and choose **Start editing**. Select the **Edit Vertex** button, and click on a 2D feature in the `Watercourse` layer. On the **3D Editor** toolbar, click on the **Sketch Properties** to see the **X, Y** coordinates of the vertices, as shown in the following screenshot:

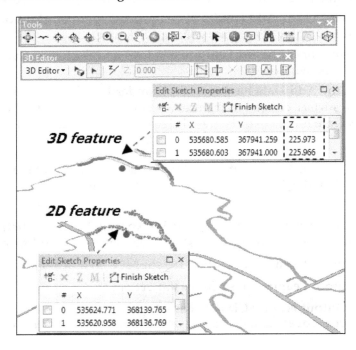

11. With the **Edit Vertex** button selected, click on the 3D feature in `Watercourse_3D` layer. In the **Edit Sketch Properties** window, notice the **X**, **Y**, and **Z** values.

12. Stop the edit session by clicking on **3D Editor | Stop Editing**. Close the **Edit Sketch Properties** window.

13. Use the **Interpolate Shape** tool again to create two 3D feature classes from the `Road` and `FishPond` feature classes. Use the `CONFLATE_ZMAX` interpolation method for the `Road` layer.

14. Use the `CONFLATE_NEAREST` interpolation method for the `FishPond` layer, and check the **Interpolate Vertices Only (optional)** option.

15. Now notice that the 3D feature classes are represented in ArcScene based on the `z`-value of their vertices, without us setting its base height.

Next, we will update `Tin_Surface` with the new 3D features:

16. In the **ArcToolbox** window, expand **3D Analyst Tools | Data management | TIN**, and double-click on the **Edit TIN** tool. Set the following parameters:

    ◦ **Input TIN**: Set as `TIN_Surface`

    ◦ **Input Feature Class**: Click on the **Browse** button, and add the following feature classes:

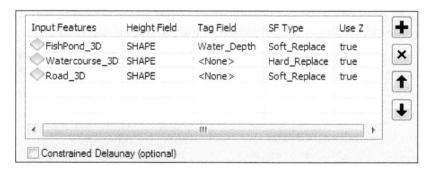

17. Click on **OK**. When complete, the updated `TIN_Surface` layer is added in **Table Of Contents**.

18. Right-click on the **FishPond_3D** layer, and choose **Zoom To Layer**. Select the **Identify** tool, and click on the fish pond surface to obtain information about the aspect, slope, and elevation values, as shown in the following screenshot:

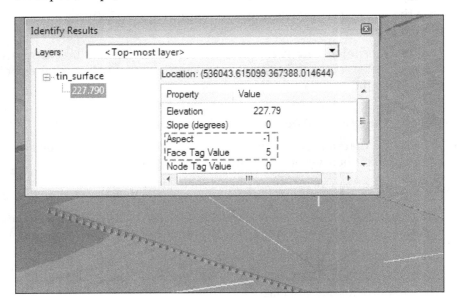

19. The `Aspect` value `-1` means that the triangle plane is horizontal. Also, notice the depth value (`5 meters`) stored in the `Face Tag Value` field.

20. Save your changes in a new scene document named `My3DFeatures.sxd` in `<drive>:\LearningArcGIS\Chapter8\3DAnalyst`.

21. Leave the ArcScene document open to learn how to calculate the area and volume of a fish pond in the next exercise.

You can find the results at: `<drive>:\LearningArcGIS\Chapter8\ Results\3DAnalyst\3DFeatures.sxd`.

# Calculating surface area and volume

In this section, we will calculate the surface area and volume of a fishing pond of length `200` meters, width `80` meters, and depth `5` meters. We will also explore a fourth way to create 3D features: by generating multipatch features between two 3D surfaces. The Esri GIS Dictionary defines the *multipatch* in the ArcGIS context, as "a type of geometry comprised of planar three-dimensional rings and triangles, which occupies discrete area or volume in three-dimensional space" (*A to Z GIS: An Illustrated Dictionary of Geographic Information Systems*, Tasha Wade and Shelly Sommer, Esri Press).

Follow these steps to start calculating the area and volume of a rectangular prism using the 3D Analyst tools in the ArcScene application:

1.  Start the ArcScene application, and open your scene document named My3DFeatures.sxd stored in the <drive>:\LearningArcGIS\ Chapter8\3DAnalyst folder.

First we will add a new field in the attribute table of the FishPond_3D layer. The field will be named Height, and will store the elevation value at the bottom of the fishing pond, which is 222 meters:

2.  Right-click on the **FishPond_3D** layer, and choose **Open Attribute Table**. From the **Table Options**, select **Add Field**. For **Name**, type Height. For **Type**, choose **Double**. Click on **OK**.

3.  Right-click on the **Height** field, choose **Field Calculator**, and then type 222. Click on **OK**.

4.  Next we will create a TIN that will cover the pond area. In the **ArcToolbox** window, expand **3D Analyst Tools | Data management | TIN**, and double-click on the **Create TIN** tool. Set the following parameters:

    °   **Output TIN**: Set as <drive>:\LearningArcGIS\Chapter8\ 3DAnalyst\TIN_fishingpond

    °   **Coordinate System (optional)**: Set as Pulkovo_1942_Adj_58_ Stereo_70

    °   **Input Feature Class**: Select the FishPond_3D layer from the drop-down list

5.  For **SF Type**, choose **Mass_Points**. For **Height Field**, select **Height**. Leave the last options unchecked, and then click on **OK** to run the tool. Once complete, TIN_fishingpond is added to **Table Of Contents**.

Since the FishPond_3D polygon feature has only four vertices, the resultant TIN will have only two triangles. Visualize the TIN_fishingpond triangles by adding the **Edges with the same symbols** rendering mode, and unchecking the default **Elevation**.

6. Resymbolize the new TIN layer, as shown in the following screenshot:

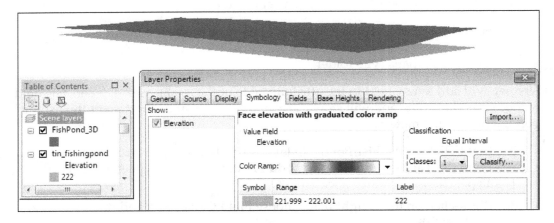

7. Next, we will create a *multipatch* feature between the two TIN surfaces. In the **ArcToolbox** window, expand **3D Analyst Tools | Triangulated Surface**, and double-click on the **Extrude Between** tool. Set the following parameters:

   ° **Input TIN**: `tin_fishingpond`

   ° **Input TIN**: `tin_surface`

   ° **Input Feature Class**: `FishPond_3D`

   ° **Output Feature Class**: `<drive>:\LearningArcGIS\`
     `Chapter8\3DAnalyst\TOPO5000.gdb\Hydrography\FishPond_`
     `MultiPatch`

8. Click on **OK** to run the tool. The polygon feature from the `FishPond_3D` layer will define the horizontal dimension, and the two TIN surfaces will constrain the vertical dimension of the multipatch block.

9. Once done, the `FishPond_MultiPatch` layer is added in **Table Of Contents**. If necessary, turn off the `tin_surface` TIN layer to see the multipatch feature or block.

10. In the **ArcToolbox** window, expand **3D Analyst Tools | 3D Features** and double-click on the **Enclose Multipath** tool. Set the following parameters:

    ° **Input Multipatch Features**: `FishPond_MultiPatch`

    ° **Output Multipatch Features Class**: `FishPond_Reservoir`

11. Click on **OK** to run the tool.

Finally, we will add the area and volume information in the attribute table of the FishPond_Reservoir layer:

12. In the **ArcToolbox** window, expand the **3D Analyst Tools | 3D Features**, and double-click on the **Add Z Information** tool. Set the following parameters:

    ° **Input Feature Class**: Set as FishPond_Reservoir

    ° **Output Property**: Check SURFACE_AREA and Volume

13. Click on **OK** to run the tool. Next, open the attribute table of the FishPond_Reservoir layer to inspect the results, as shown in the following screenshot:

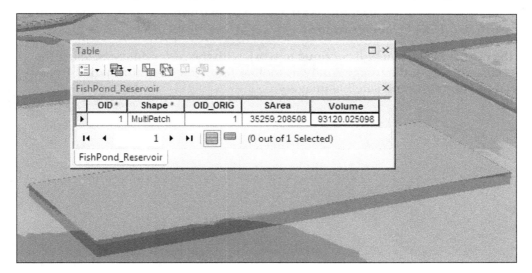

 The SAreas field stores the surface area of the rectangular prism.

14. Save your scene document as MyAreaVolume.sxd in <drive>:\ LearningArcGIS\Chapter8\3DAnalyst.

You can find the results at: <drive>:\LearningArcGIS\Chapter8\ Results\3DAnalyst\AreaVolume.sxd.

 For more exercises using the Spatial Analyst and 3D Analyst extensions, please refer to *Chapter 9, Working with Spatial Analyst,* and *Chapter 10, Working with 3D Analyst* of *ArcGIS for Desktop Cookbook, Daniela Cristiana Docan, Packt Publishing, 2015*: https://www.packtpub.com/ application-development/arcgis-desktop-cookbook.

# Summary

In this chapter, you learned how to create, display, and analyze cell-based datasets. You performed a binary site selection analysis using the ArcMap application and Spatial Analyst extension. You also learnt how to derive different raster surfaces from an elevation raster, such as hillshade, slope, and aspect. Within the least-cost path analysis, you learnt how to convert vector data to raster data, and to reclassify two rasters, that is, store categorical data (`Slope_Elev` raster layer) and continuous data (`LandUse_Raster` raster layer).

In the second part of the chapter, you created 3D features and 3D surfaces. Using the 3D data, you built a 3D multipatch feature by extruding a 3D polygon feature between two TIN elevation surfaces.

In the next chapter, you will continue to work with the raster datasets, using the specialized tools of the **Image Analysis** toolbar for extracting information from aerial photography and satellite imagery.

# 9
# Working with Aerial and Satellite Imagery

In this chapter, we will work with aerial and satellite imagery. We will explore the georeferencing workflow for rasters and three image-processing functions to extract information from the satellite imagery.

By the end of this chapter, you will learn about the following topics:

- Georeferencing aerial imagery
- Using ArcGIS tools to access and enhance the display of the images for visual interpretation
- Using the Image Analysis toolbar to extract information from satellite imagery

## Using aerial imagery

Aerial imagery or a photograph is a nadir/vertical image of the earth's surface using a high-resolution digital camera and taken from a manned or unmanned airborne vehicle. The geometric distortions of the vertical photograph, which are imposed by the camera tilt, relief displacements, and scale variations throughout the photograph, can be removed through the differential rectification or orthorectification process. An orthorectified image is called an **orthophoto**. In ArcGIS, you can use the orthophotos for the following:

- As basemaps or orthophotomaps
- As background to update a vector dataset using onscreen digitizing
- To collect vectors and attribute data through the orthophoto interpretation (manual or automated process)

# Georeferencing an orthorectified image

The Esri GIS Dictionary defines georeferencing as "aligning an image to a known coordinate system so that it can be viewed, queried, and analyzed with other geographic data." (*A to Z GIS: An Illustrated Dictionary of Geographic Information Systems, Tasha Wade* and *Shelly Sommer, Esri Press*).

To georeference a raster dataset in ArcMap, you need a *reference layer* that covers the same area and is assigned a specific Earth-related coordinate system. A reference layer may be a vector dataset, an existing raster dataset, a basemap from ArcGIS Online, or a text file that contains links between the points on the raster and their true location on the ground (ground control points).

Follow these steps to manually georeference an aerial image in the ArcMap application:

1.  Start the ArcMap application and open a blank map document.

2.  Open the **Catalog** window and connect to the Chapter9 folder using the **Connect To Folder** button. Expand the AerialImage_Orthophoto folder. This folder contains a TIFF raster dataset, representing an orthorectified photograph that was acquired in 2010. It has 0.5 meter spatial resolution. This is the orthophoto that you will be georeferencing.

3.  Right-click on the raster dataset and choose **Properties**. Inspect the information about the data source, the number of columns and rows, the number of bands, the cell size, the raster format, the extent, the spatial reference, or the statistics. Note that **Spatial Reference** is <Undefined>. Click on **OK**.

4.  In the **Catalog** window, select the existing map document named **Georeferencing.mxd** and drag it in the map display. The map document contains one raster layer representing the orthophoto named Aerial_Image_Orthorectified.tif, and one feature layer representing the target or destination coordinate system, as shown in the following screenshot:

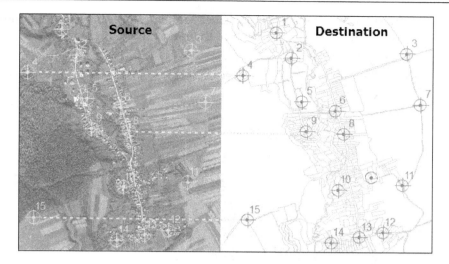

5.  The source of the LandUse (Destination) layer is <drive>:\
    LearningArcGIS\Chapter8\SpatialAnalyst\TOPO5000.gdb.

6.  In the map document, there are two **Layers Overview** windows opened,
    which may be overlapping. If necessary, click on the first **Layers Overview**
    window and move its position slightly so that you can see both overview
    windows.

7.  In ArcMap, after applying all georeferencing operations, the resulting
    orthophoto will automatically inherit the data frame's coordinate system.
    Check the data frame's coordinate system. It is set as Pulkovo_1942_Adj_58_
    Stereo_70.

8.  From the **Customize** menu, select **Georeferencing** from **Toolbars**.

9.  On the **Georeferencing** toolbar, make sure that Aerial_Image_
    Orthorectified.tif is selected as the georeferencing layer. Click on
    **View Linked Table** to add the links.

10. Next, from the **Bookmark** menu, select **Manage Bookmarks**. Move the two
    windows so that you can see both of them and the map, as shown in the
    following screenshot:

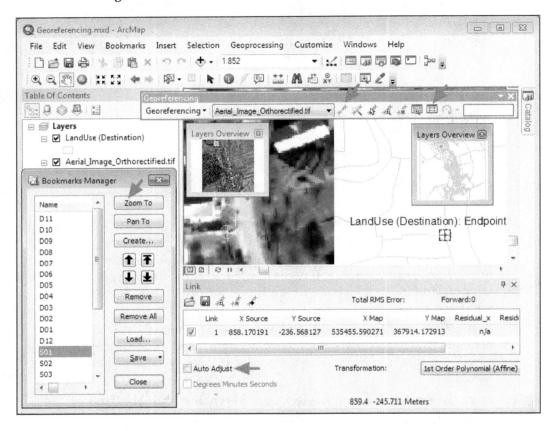

11. First, in the **Link** window, deselect **Auto Adjust** and make sure that the
    **Transformation** method is set to **1st Order Polynomial (Affine)**.

    To perform an affine transformation method, you need at least four control
    points (which are evenly distributed across the study area) with known
    coordinates in both coordinate systems (raster coordinate space and real-world
    coordinate system).

    To obtain a small value of the total **root mean square error (RMSE)**, we
    will use more than four control points. For convenience, we chose and
    marked 15 points on `Aerial_Image_Orthorectified.tif` and on the
    `LandUse` referenced layer for you. The control points are evenly distributed
    throughout the whole area and represent punctual features, such as visible
    fence corners, road intersections, or stable parcel corners.

12. Second, on the **Georeferencing** toolbar, click on the **Add Control Points** button.

13. From the **Bookmark** menu, select the **S01** bookmark and click on the **Zoom To** button to see the position of the first point on the orthophoto.

14. Click on the point marked with a green circle to add the first point of the link.

15. Then, select the **D01** bookmark in the **Bookmarks Manager** window and click on the **Zoom To** button to see the destination point on the LandUse layer.

16. Click on the marked point to add the second point of the link (the control point).

17. In the **Link** window, the record of the first link is added. The coordinates of the point that you clicked on in the orthophoto are stored in the **X Source** and **Y Source** fields. The destination point coordinates are stored in the **X Map** and **Y Map** fields.

18. Continue to add fourteen more control points, as shown in the following screenshot:

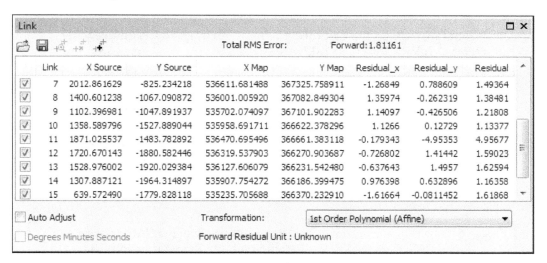

| Link | X Source | Y Source | X Map | Y Map | Residual_x | Residual_y | Residual |
|---|---|---|---|---|---|---|---|
| 7 | 2012.861629 | -825.234218 | 536611.681488 | 367325.758911 | -1.26849 | 0.788609 | 1.49364 |
| 8 | 1400.601238 | -1067.090872 | 536001.005920 | 367082.849304 | 1.35974 | -0.262319 | 1.38481 |
| 9 | 1102.396981 | -1047.891937 | 535702.074097 | 367101.902283 | 1.14097 | -0.426506 | 1.21808 |
| 10 | 1358.589796 | -1527.889044 | 535958.691711 | 366622.378296 | 1.1266 | 0.12729 | 1.13377 |
| 11 | 1871.025537 | -1483.782892 | 536470.695496 | 366661.383118 | -0.179343 | -4.95353 | 4.95677 |
| 12 | 1720.670143 | -1880.582446 | 536319.537903 | 366270.903687 | -0.726802 | 1.41442 | 1.59023 |
| 13 | 1528.976002 | -1920.029384 | 536127.606079 | 366231.542480 | -0.637643 | 1.4957 | 1.62594 |
| 14 | 1307.887121 | -1964.314897 | 535907.754272 | 366186.399475 | 0.976398 | 0.632896 | 1.16358 |
| 15 | 639.572490 | -1779.828118 | 535235.705688 | 366370.232910 | -1.61664 | -0.0811452 | 1.61868 |

Total RMS Error: Forward: 1.81161

Auto Adjust     Transformation: 1st Order Polynomial (Affine)

Degrees Minutes Seconds     Forward Residual Unit : Unknown

19. Your coordinate values from the records may differ slightly from the ones in the previous screenshot. If you make a mistake while adding a link or it has high residual error values, select the wrong link record and click on the **Delete Link** button or press the *Delete* key. Then add the correct link.

> If you didn't succeed in identifying all 15 points, you can load the correct links from the Links.txt file stored at <drive>:\
> LearningArcGIS\Chapter9\AerialImage_Orthophoto.

20. When finished, on the **Georeferencing** toolbar, navigate to **Georeferencing | Update Display**. On the map, `Aerial_Image_Orthorectified.tif` is displayed temporarily in the real-world coordinate system that you just georeferenced it to using the links.

21. In the **Link** window, the value of **Total RMS Error** is displayed. Click on the **Save** icon to save the links from the list as a text file. You can reuse this file if you want to repeat the georeferencing process without repeating the previous steps.

22. To save the `Aerial_Image_Orthorectified.tif` raster layer as a permanent georeferenced raster on the disk, on the **Georeferencing** toolbar, navigate to **Georeferencing | Rectify**.

23. You can define the cell size, the **NoData** value, the resampling method, the output location, the file format and name, and the compression type. For **Resample Type**, choose **Bilinear Interpolation (for continuous data)**. For **Output Location**, select `<drive>:\LearningArcGIS\Chapter9\AerialImage_Orthophoto` folder. For **Name**, type `Orthophoto_Stereo70`.

24. Accept the default values for all other parameters. Click on **Save**.

25. When finished, on the **Georeferencing** toolbar, go to **Georeferencing | Delete Links**. Close the **Link** window.

26. From the **File** menu, select **New**. You do not need to save the changes in the `Georeferencing.mxd` map document because the georeferenced output raster was saved on disk when you rectified the `Aerial_Image_Orthorectified.tif` raster layer.

27. Add the `Orthophoto_Stereo70` output raster in **Table Of Contents**. Inspect the raster properties. You can add the `LandUse` feature class results from `<drive>:\LearningArcGIS\Chapter8\SpatialAnalyst\TOPO5000.gdb`. Inspect the results.

The **Georeferencing** toolbar can also work with the CAD datasets (DWG, DXF and DGN) through **Similarity Transformation**. This transformation method requires two transformation links. The **Update Georeferencing** tool will generate a World file (`*.WLD`). As ArcGIS for Desktop does not allow you to edit or change the CAD dataset, ArcMap will read the georeferencing control points from the WLD file and will apply translations on X and Y, a rotation, and a scale to the CAD dataset so that it displays in a real-world coordinate system.

If you want to change the original source CAD data, export it as a feature class to a geodatabase and use the **Spatial Adjustment** tools that you have already learned in the *Using spatial adjustment* section in *Chapter 5, Creating and Editing Data*.

28. Close ArcMap without saving your changes.

You can find the results at `<drive>:\LearningArcGIS\Chapter9\Results\AerialImage_Orthophoto\Georeferencing.mxd`.

# Using satellite imagery

In ArcGIS, you can use the satellite imagery, as follows:

- As a basemap and background for the onscreen digitizing process
- As a source of attribute data through the visual interpretation process
- For advanced analysis using the image processing tools (for example, Normalized Difference Vegetation Index-NDVI)

The source of the **Landsat 7 ETM+** satellite imagery that is used in this chapter is the Global Land Cover Facility, `www.landcover.org`. The Landsat data files were created by the U.S. Geological Survey (USGS), and stored in the **Geographic Tagged Image-File Format (GeoTIFF)**. (*NASA Landsat Program, 2007, Landsat ETM+ scene L71183029_02920070727, L1G, USGS, Sioux Falls, 07/27/2007*)

A satellite image is a multispectral image with three or more spectral bands. A spectral band captures the spectral signature of the objects within a range of frequencies across the electromagnetic spectrum. A Landsat 7 ETM+ image includes *three* visible bands (blue, green, and red), *five* infrared bands (near, mid, and thermal IR), and *one* panchromatic band. (Source: `http://glcf.umd.edu/data/landsat/`)

The multiband imagery is displayed in ArcMap as a three-band color image using the additive primary colors that are used by the computer screen: Red, Green, and Blue. Each spectral band of the Landsat image is displayed with one of the three colors of the RGB composite render in ArcMap.

For more information about the dataset, use the Notepad application to open and read the `README.GTF` file from `<drive>:\LearningArcGIS\Chapter9\SatelliteImagery`.

This section does not undertake covering all fundamental concepts of satellite imagery data because it is beyond the scope of this book. Please refer to `http://glcf.umd.edu/data/landsat`, and read the **Technical Guide** and **File Format Guide**.

# Accessing imagery

There are three main ways to store and access imagery in ArcGIS for Desktop:

- As a raster dataset stored in a filesystem raster or in a geodatabase
- Through a mosaic dataset in a geodatabase
- Using an image service

We call a raster dataset any single-band raster (continuous or discrete) or multiband raster, such as aerial and satellite imagery, which is stored in an individual file on disk or in a geodatabase.

In this section, we will create a mosaic dataset and organize imagery inside it. Then, we will access satellite imagery through a web service.

A mosaic dataset is a collection of raster datasets inside the geodatabase. A mosaic dataset manages collections of *rasters* (for example, scanned map, or thematic raster) or *imagery* (for example, aerial, and satellite) without loading the source pixels of the raster file. A mosaic dataset just references the raster datasets that are stored inside or outside a geodatabase. A mosaic dataset has the following capabilities:

- Dynamically mosaicking the overlapping raster datasets
- Different image processing functions that are applied on-the-fly in ArcMap
- Managing the metadata of the mosaicked image

To enhance the display speed of the mosaic dataset at different scales and reduce the CPU (computer central processing unit) usage, ArcMap builds reduced resolution copies of the original raster datasets named **pyramids** (optional) and **overviews**. Both are based on the original pixel size of the source raster datasets.

Pyramids are created for raster datasets and have a downsampling factor of $2x$. For a source raster dataset with 15-meter cell size, ArcMap will build two reduced resolution copies of the raster. The first level of the pyramid will have a 30-meter resolution, and the second level of pyramid will have a 60-meter resolution.

Overviews are analogous to pyramids, but they are created for a mosaic dataset with a downsampling factor of $3x$. In the next exercise, we will provide you with more details about the spatial resolution of the overviews.

For more information about the cell size ranges for the raster datasets and mosaic datasets, please refer to the ArcGIS Resource Center at http://resources.arcgis.com/en/help/.
Navigate to **Desktop (ArcMap): 10.4 | Manage Data | Data types | Raster and images | Building and managing a raster database | Mosaic datasets**

Follow these steps to access the imagery from a mosaic dataset and an image service:

1. Start the ArcMap application and open a blank map document.

2. Open the **Catalog** window and expand the `MosaicDataset` folder. The folder contains three TIFF images or raster datasets.

3. Right-click on the **landsat_7_bands_UTM35N2.tif** raster dataset and choose **Properties**. There are six sections of the raster properties that provide you with information about the data source, the number of columns and rows, the number of bands, the cell size, the raster format, the extent, the spatial reference, or the statistics. Click on **OK**.

4. Explore **Raster Dataset Properties** for all three TIFF images. Please note that the raster datasets have two different resolutions: the `landsat_7_bands_UTM35N2` and `Landsat_7_bands_UTM35N3` images have a resolution of 30-meter cell size, and the `Landsat_8_bands_UTM35N1` image has a resolution of 15-meter cell size.

5. On the **Standard** toolbar, click on the **Add Data** button. Add all three TIFF images from the `<drive>:\LearningArcGIS\Chapter9\MosaicDataset` folder. In the **Create Pyramids** window, click on **Yes** to build pyramids for the raster datasets on the map.

6. On the **Tools** toolbar, click on the **Full Extent** button to display the full extent of the satellite imagery.

7. With the *Ctrl* key pressed, deselect the first raster layer to deselect all layers at once. Inspect the imagery by checking the layers one by one.

8. When finished, deselect all raster layers in **Table Of Contents**.

9. To create a mosaic dataset, we need a geodatabase. Right-click on the `MosaicDataset` folder, and navigate to **New | File Geodatabase**. Change the default name of the geodatabase to `Europe.gdb`.

Next, we will create an empty mosaic dataset within the `Europe.gdb` geodatabase, as shown in the following screenshot:

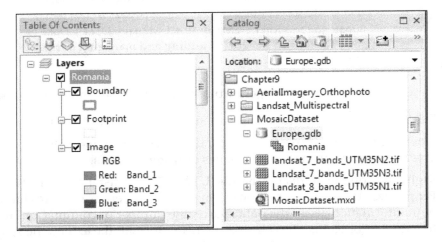

10. Right-click on `Europe.gdb`, and go to **New | Mosaic Dataset**. Set the following parameters:

    ○ **Output location**: This should be `D:\LearningArcGIS\Chapter9\Europe.gdb`

    ○ **Mosaic Dataset Name**: This should be `Romania`

    ○ **Coordinate System**: This should be `WGS 1984 UTM Zone 35N` from **Projected Coordinate Systems | UTM | WGS 1984 | Northern Hemisphere**

11. Accept the default values for the rest of the parameters. Click on **OK**.

12. The empty `Romania` mosaic dataset is added in the `Europe.gdb` geodatabase and also to **Table Of Contents**.

The `Romania` mosaic dataset, contains three layers:

- The `Boundary` layer displays the boundary of the entire area covered by the collection of the imagery

- The `Footprint` raster layer displays the extent of each raster dataset in the mosaic dataset, including the extent of the overview rasters

- The `Image` layer controls the order of the overlapping rasters in the mosaicked image using different mosaicking methods, such as **North-West, Closest to Center**, or **By Attributes**

Next, we will add all three pieces of imagery to the `Romania` mosaic dataset, as shown in the following screenshot:

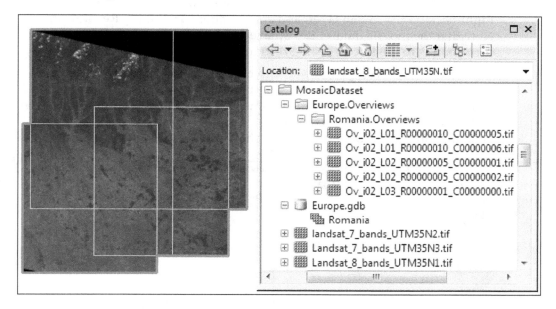

Follow these steps to add the satellite imagery in the `Romania` mosaic dataset:

1. In the **Catalog** window, right-click on the **Romania** mosaic dataset, and choose **Add Rasters**. Set the following parameters:

   ◦ **Input Data**: This should be `Dataset`

   ◦ **Source**: This should be according to the following screenshot:

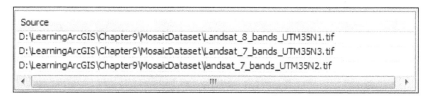

   ◦ Check **Update Overviews (optional)**

2. Accept the default values for the rest of the parameters. Click on **OK**.

3. The overview rasters are stored in a new folder named `Romania.Overviews`. In the **Catalog** window, select the **MosaicDataset** folder and press the *F5* key to refresh its content. Expand the `Europe.Overviews\Romania.Overviews` folder and inspect the properties of overview rasters.

Next, we will select the footprint of the `Landsat_7_bands_UTM35N3` raster dataset:

4. In **Table Of Contents**, make sure that the **Boundary** layer is deselected. Right-click on the **Image** layer, and choose **Zoom To Source Resolution**. Note that the mosaic dataset is displayed at scale `1:56,693`, which corresponds to the lowest spatial resolution of the source raster datasets.

5. Right-click on the **Footprint** layer, and choose **Open Attribute Table**. The **Footprint** attribute table displays the satellite imagery and the overviews, along with their attributes, such as name, minimum and maximum cell sizes, and raster categories. Inspect the `LowPS`, `HighPS`, `MinPS`, and `MaxPS` values. Note the three levels of overviews, as shown in the following screenshot:

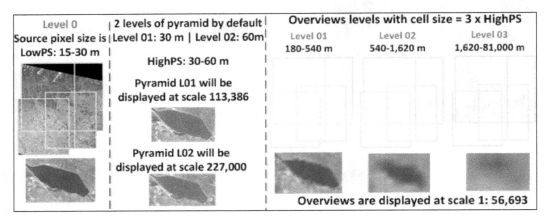

6. In the **Table** window, select `Landsat_7_bands_UTM35N3`. Note the extent of the selected raster is highlighted in blue on the map. Click the **<Raster>** cell within the **Raster** column to open the **Romania: 2** window, as shown in the following screenshot:

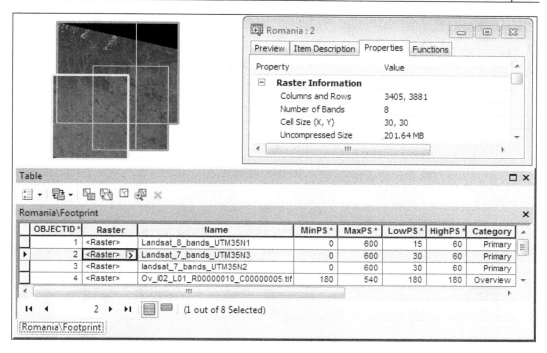

7. Inspect the properties of the **Landsat_7_bands_UTM35N3** image.

8. When finished, close the **Romania: 2** window. To deselect the raster, click on **Clear Selection**. Keep the **Footprint** table window open.

9. Let's try a second option of selecting a raster. With the **Select Features** tool, try to select the **Landsat_7_bands_UTM35N3** image on the map. As the footprints of the satellite imagery are overlapping with the overviews, you have selected four records in the table.

10. In **Table Of Contents**, right-click on the **Footprint** layer, and navigate to **Selection | Reselect Only Primary Rasters**. Now, you have only the primary raster selected.

11. In **Table Of Contents**, right-click again on the **Footprint** layer, and navigate to **Selection | Lock To Selected Raster**. Now, only the Landsat_7_bands_ UTM35N3 raster is visible on the map. Close the **Romania\Footprint Table** window.

Next, we will once again display all raster datasets in the Mosaic Dataset:

1.  In the **Table Of Contents**, right-click on the **Image** layer, and navigate to **Properties | Mosaic**. Move the **Layer Properties** window so that you can see both, the window and the mosaic dataset, on the map.

2.  Note that **Mosaic Method** is set to **Lock Raster**. This is because you have worked with the **Lock To Selected Raster** tool. In the drop-down list, select **None**, and click on **OK**. All raster datasets are displayed.

    The ArcMap provides many mosaic methods to control how the overlapping rasters and overviews appear in the mosaic dataset. Next, we will display the highest-resolution image on top of the other imagery in the Romania mosaic dataset, as shown in the following screenshot:

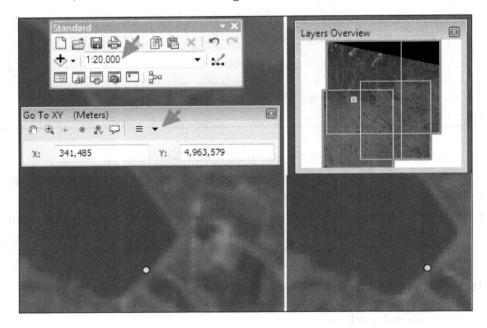

3.  On the **Standard** toolbar, click on the **Go To XY** tool. For the **Units** choose **Meters**. For **X**, type 341,485, and for **Y**, type 4,963,579. Press the *Enter* key.

4. On the **Go To XY** tool, click on the **Zoom To** and **Flash** tools to identify the point position on the map. Then, set the view scale to 20,000 to see the whole man-made lake (reservoir).

5. In **Table Of Contents**, right-click on the **Image** layer, and navigate to **Properties | Mosaic**. Move the **Layer Properties** window so that you can see both, the window and the mosaic dataset, on the map.

6. Set **Mosaic Method** to **By Attribute**, and then set **Order Field** to **LowPS**. Check **Order Ascending**, and click on **Apply**. Inspect the results on the map display. Note the resolution changes on the map. The image with the lowest cell size appears first in the mosaic dataset.

You can modify the mosaic method from the following:
- The Image **Layer Properties** dialog window: the changes will affect only the display of the raster layer in ArcMap without altering the mosaic dataset properties
- The **Mosaic Dataset Properties** dialog box in the **Catalog** window: the changes will permanently modify the mosaic dataset

7. Close the **Layer Properties** window. Zoom to the full extent.

Next, we will change the displaying of the image using the Landsat 3, 2, 1 combination for a more natural display, as shown in the following screenshot:

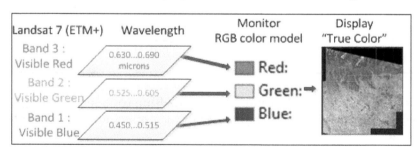

Follow these steps to display the mosaic dataset as a "true color" image:

1.  In **Table Of Contents**, right-click on the **Image** layer, and navigate to **Properties | Symbology**. Click on the down arrow, next to the **Red** row, and select **Band_3**, as shown in the following screenshot:

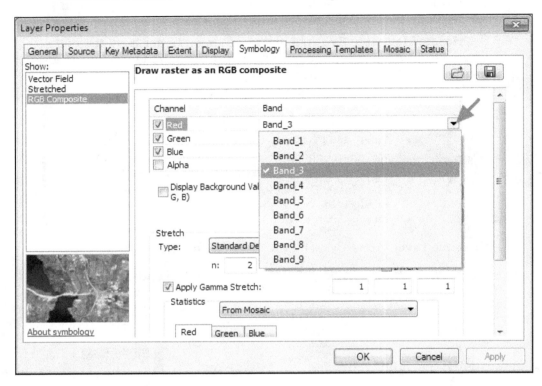

2.  For the **Green** row, select **Band_2**. For the **Blue** row, select **Band_1**. Accept the default values for the rest of the parameters. Click on **OK**. Inspect the results.

In ArcMap, you can change the default band combination for a raster layer by navigating to **Customize | ArcMap Options | Raster**, and selecting **Raster layer**, and setting the bands in the **Default RGB Band Combinations** section.

The forest is dark green. The water is dark blue or green. The bare soil has different shades of brown, and the crops are light green. The buildings are silver, gray, or red, and the roads are gray. Snow and clouds are white.

Next, we will access the Landsat 7 satellite imagery through the ArcGIS Online Landsat image service, as shown in the following screenshot:

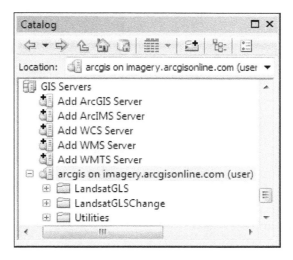

Follow these steps to access the image service:

1. In the **Catalog** window, double-click on the **Add ArcGIS Server** under the **GIS Servers** node to connect to an image service. Click **Next**. Next to **Server URL**, type `http://imagery.arcgisonline.com/arcgis/services`. Click on **Finish**.

2. Expand the **LandsatGLS** folder. Right-click on the **TM_Multispectral_2010** image service and select **Item Description**. Inspect the metadata information.

   The Landsat multispectral image has a spatial resolution of 30 meters, and the projected coordinate system is `WGS_1984_Web_Mercator_Auxiliary_Sphere`:

3. When finished, close the **Item Description** window.

4. Select and drag the **TM_Multispectral_2010** image service on the map display. Move the **LandsatGLS\TM_Multispectral_2010** layer below the **Romania** mosaic dataset, as shown in the following screenshot:

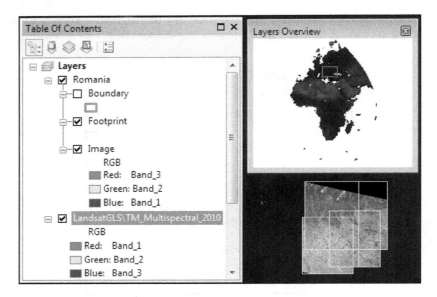

In the preceding screenshot, the red rectangle in the **Layers Overview** window represents the extent of the map display. To display your imagery at different scales, you can add the **Overview** window from the **Windows** menu:

5. Right-click on the **Layers Overview** window and select **Properties**. From the **Reference layer** drop-down list, select the **LandsatGLS\ TM_Multispectral_2010** layer. Click on **OK** to save changes and close the **Overview Properties** window.

6. Use the zoom tools on the **Standard** toolbar to inspect the results. When finished, save the changes in the map document as AccessingImagery.mxd to <drive>:\LearningArcGIS\Chapter9\MosaicDataset.

7. Leave the AccessingImagery.mxd map document open to enhance the display of the TM_Multispectral_2010 layer in the next exercise.

You can find the results at <drive>:\LearningArcGIS\Chapter9\Results\ MosaicDataset\AccessingImagery.mxd.

# Displaying multispectral imagery

In this section, we will learn how to enhance the displaying (rendering) of an image service and mosaic dataset in ArcMap for visual interpretation.

The **Image Analysis** toolbar provides you with different tools to enhance the display of the raster dataset in ArcMap without altering the cell values of the original source data, such as the following:

- Calculating statistics and building histograms based on the pixel values, which are defined by the bit depth of the raster (for example, 8-bit means that pixel values may lie between 0 and 255), as shown in the following screenshot:

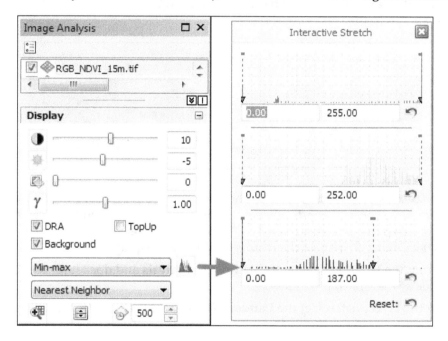

- Applying different stretch methods to spread the pixel values along the histogram in order to improve the contrast of the image; a stretch method cannot be applied without statistics or histograms

- Resampling the cell values of image; for example, **Nearest neighbor** and **Majority** are used to display the discrete thematic rasters; **Bilinear interpolation** and **Cubic convolution** are used to display aerial and satellite imagery or continuous thematic rasters (Source: www.esri.com)

- Controlling the brightness, contrast, gamma, and transparency

Follow these steps to enhance the display of satellite imagery in ArcMap:

1. Start the ArcMap application and open your map document named
   `AccessingImagery.mxd` from `<drive>:\LearningArcGIS\Chapter9\`
   `MosaicDataset`.

2. In **Table Of Contents**, deselect all layers except the **LandsatGLS\TM_Multispectral_2010** image service. Right-click on the image service and select **Zoom To Source Resolution**.

We would like to perform a visual analysis of the road infrastructure. To distinguish the highway features on the `LandsatGLS\TM_Multispectral_2010` layer more easily, we will change the default band combination to Landsat `4`, `3`, `2` or "False color", as shown in the following screenshot:

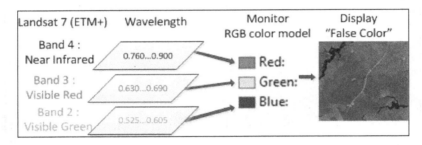

3. In **Table Of Contents**, click on the **Red: Band_1** red square to see the available spectral bands and select **Band_4**. Then, click on the **Green: Band_2** green square and select **Band_3**. Click on the **Blue: Band_3** blue square and select **Band_2**. Inspect the results on the map display.

4. From the **Windows** menu, select **Image Analysis** and dock it on the right-hand side of the ArcMap window.

5. In the layers list, select the **LandsatGLS\TM_Multispectral_2010** layer, as shown in the following screenshot:

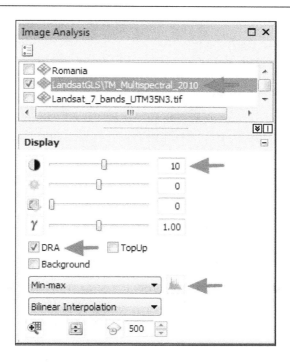

6. In the **Display** section, below the **Background** box, select **Min-max** stretch type from the first drop-down list and click on **Yes** to calculate statistics and build the histogram.

7. Select the **DRA** option that refers to dynamic range adjustment of the contrast and brightness characteristics only for the values of pixels in the current display extent, and not for the pixel values in the entire LandsatGLS\ TM_Multispectral_2010 image service (Source: http://desktop.arcgis. com).

8. Finally, set **Contrast** to 10. Make sure that the **Brightness** and **Transparency** fields have value 0, and **Gamma** is 1. Inspect the results in the map display.

The forest ranges from red to dark red. The water is dark blue, and if it contains sediments, the color ranges from blue to cyan. The crops range from the colors pink to bright red, depending on their states of growth and health. The bare soil is blue. The buildings range from the colors blue to gray, depending on the roof material, and the roads are cyan. Snow and clouds are white or cyan:

9.  Let's inspect a specific area. From the **Bookmarks** menu, select **Manage Bookmarks**. Click on the **Load** button and select the ImageProcessing.dat file from <drive>:\LearningArcGIS\Chapter9\MosaicDataset. Click on **Open**.

10. Select the **Highway** bookmark, and click on the **Zoom To** button. Click on **Close**.

11. In **Table Of Contents**, right-click on the **LandsatGLS\TM_Multispectral_2010** layer, and select **Zoom To Source Resolution**. The highway appears in cyan. You may notice the bridges that cross the highway.

The Romania mosaic dataset contains satellite imagery from 2007 and the LandsatGLS\ TM_Multispectral_2010 layer contains satellite imagery from 2010. Let's check the evolution of highway construction between 2007 and 2010, as shown in the following screenshot:

12. In **Table Of Contents**, check the Romania mosaic dataset. Change the default 1, 2, 3 band combination of the Image layer to the 4, 3, 2 bands, as you have learned in the previous steps.

13. In the **Image Analysis** window, select the **Romania** layer. Then, in the **Display** section, select the **Std-dev** stretch type and click on **Yes** to calculate statistics and build the histogram.

14. Make sure that the **Romania** layer is still selected. In the **Display** section, click on the **Swipe Layer** tool. On the map display, the mouse pointer is displayed as a black arrow. Drag down to swipe between the Romania/Image and LandsatGLS\TM_Multispectral_2010 layers.

15. Next to the **Swipe Layer** tool, select the **Flicker layer** tool to interactively see the difference between the layers. To turn off **Flicker layer**, click on it again.

16. Save the changes in your map document as DisplayingImagery.mxd to <drive>:\LearningArcGIS\Chapter9\MosaicDataset. Close ArcMap.

You can find the results at <drive>:\LearningArcGIS\Chapter9\Results\MosaicDataset\DisplayingImagery.mxd.

 For more information about satellite image interpretation, please refer to: http://earthobservatory.nasa.gov/Features/ColorImage/.

# Image processing

In this section, we will explore the main processing functions, such as composing spectral bands, pan-sharpening, and NDVI, using the Image Analysis window.

In this section, we will explore the most commonly used image processing functions:

- **Composite bands**: This creates a single multispectral image from multiple single-band images

- **Pan Sharpen**: This uses a high-resolution panchromatic (single-band) image and a lower-resolution multiband image to create a multiband image with the spatial resolution of the panchromatic image

- **NDVI** (Normalized Difference Vegetation Index): This maps vegetation using the difference between the visible red band (visible red light is absorbed by the vegetation) and the near-infrared band (reflected by vegetation). You can use the NDVI map to detect the presence or absence of vegetation or the type and health of vegetation

Follow these steps to start working with image processing functions:

1.  Start the ArcMap application and open a blank map document. Open the **Catalog** window and expand the `<drive>:\LearningArcGIS\Chapter9\ SatelliteImagery\ L71183029_02920070727.ETM` folder. This folder contains a set of nine TIFF imagery representing eight spectral bands, including a thermal band divided into two bands and one panchromatic band (`http://landsat.usgs.gov/about_landsat7.php`), as shown in the following screenshot:

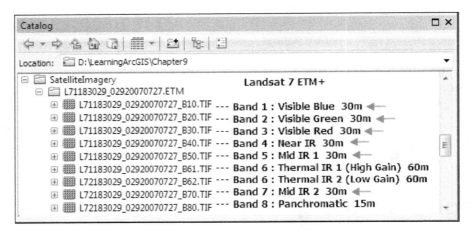

2.  On the **Standard** toolbar, click on the **Add Data** button. Select and add the following imagery, from `B01` to `B50`, and `B70`.

3.  In the **Create Pyramids** window, click on **Yes** to build pyramids for the raster datasets on the map. All six single-bands rasters are displayed as grayscale raster layers. Every image represents the same area and it stores a single band.

4.  In **Table Of Contents**, with the *Ctrl* key pressed, click on the expansion control small icon (-) from the left side of the first raster layer to collapse all layers at once. Inspect the properties of the satellite imagery.

5.  When finished, click on the **Full Extent** button to display the full extent of the satellite imagery.

    Even if we worked with the multispectral imagery in the previous exercise, we will create our own *multispectral image* or *composite band image* using all six single-band raster datasets. This *multiband image* will display all six spectral bands as an RGB composite in ArcMap, as shown in the following screenshot:

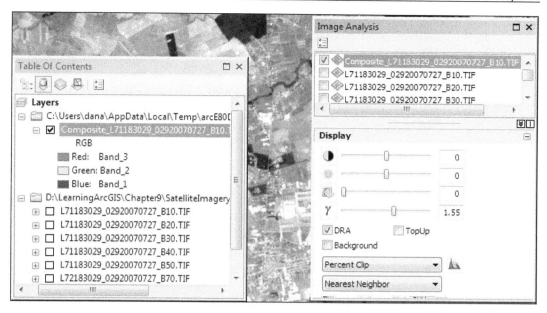

6. Open the **Image Analysis** window and dock it on the right-hand side of the ArcMap window. In the layers list of the **Image Analysis** window, right-click on the first image named `L71183029_02920070727_B10.TIF` and navigate to **Properties | source**.

7. In **the Raster Information** notice section, **Number of bands** subsection, that the raster dataset has only one band, representing *Band 1: Visible Blue* and a spatial resolution of 30 meters. Click on **OK**.

8. Use the *Shift* key to select all the layers in the **Layer** list. Then, in the **Processing** section, click on the **Composite Bands** button.

9. A new temporary raster dataset named `Composite_L71183029_02920070727_B10.TIF` is added on-the-fly in **Table Of Contents**.

> All output rasters, which result from applying the image processing functions from the **Image Analysis** toolbar, are considered temporary raster layers in your map document. In order to make a raster layer a permanent raster dataset, select it in the **Layer** list from **Image Analysis**, and then click on the **Export** button from the **Processing** section.

10. In **Table Of Contents**, turn off all layers except the new composite raster layer.

11. In the **Image Analysis** window, right-click on the **Composite_L71183029_02920070727_B10.TIF** layer, and select **Functions** from **Properties**. Move the **Layer Properties** window so that you can see the map display and the **Image Analysis** window.

12. The default function named **Composite Band Function** was automatically applied to combine all six bands in a multiband raster.

13. Click on the **Symbology** tab. Change the display of the spectral band combination 1, 2, 3 to the True Color: **Red: Band_3**, **Green: Band_2**, and **Blue: Band_1**.

14. For **Type** from **Stretch**, select **Percent Clip** to eliminate the extreme pixel values in the histogram and obtain a bright image. Click on **Apply**. Note that the stretch type is updated in the **Image Analysis** window.

15. In **Layer Properties** window, click on the **Symbology** tab, and then select the **From Current Display Extent** option from the **Statistics** drop-down list. Click on **Apply**. Note that the **DRA** display option is selected in the **Image Analysis** window.

16. Click on the **Source** tab. Note that the raster has a 30-meter cell size resolution, and this is a temporary one. Click on **OK** to close the **Layer Properties** window. Inspect the results.

Next, we will increase the resolution of the Composite_L71183029_02920070727_B10.TIF composite image using the pan-sharpening technique. The result will be a *high-resolution multiband image* or a *pan-sharpened multispectral image* with a spatial resolution of 15 meters, as shown in the following screenshot:

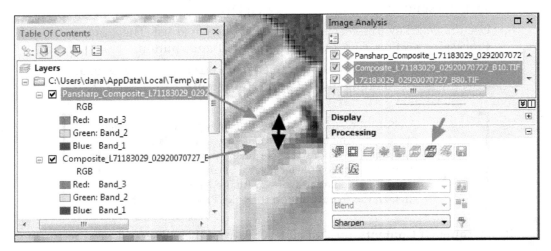

Follow these steps to start working with the pan-sharpening function:

1.  In the **Catalog** window, select and drag the `L72183029_02920070727_B80.TIF` panchromatic image on the map display.

2.  In **Table Of Contents**, drag the panachromatic image below the temporary composite image. Inspect the properties of the panchromatic image and note that it has a higher resolution than the temporary composite image.

3.  In the **Image Analysis** window, select both composite and panchromatic imagery. Then, in the **Processing** section, click on the **Pan Sharpen** button.

4.  A new pan-sharpened image named `Pansharp_Composite_L71183029_02920070727_B10.TIF` is added to **Table Of Contents**.

5.  Open its **Layer Properties** window and inspect the **Source** and the **Functions** sections. Then, click on **Symbology** and set the display characteristics, as shown in steps 11 to 13 of the previous list. Click on **OK** to close the **Layer Properties** window.

6.  Let's explore a specific area on the map display. In the **Catalog** window, select the `AreaOInterest.lyr` layer file from the `<drive>:\LearningArcGIS\Chapter9\SatelliteImagery` folder. Drag this on the map display.

7.  In Table **Of Contents**, right-click on the **AreaOfInterest** layer, and select **Zoom To Layer**. Then, right-click on the pan-sharpened image and select **Zoom To Raster Resolution**.

8.  Inspect the difference in resolution using **Swipe layer** from the **Image** Analysis window.

In the last part of this exercise, we will calculate and represent the vegetation index named Normalized Difference Vegetation Index (NDVI) only for the area covered by the AreaOfInterest layer, as shown in the following screenshot:

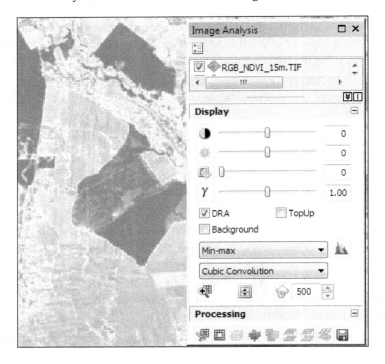

9. If necessary, right-click on the **AreaOfInterest** layer and select **Zoom To Layer**. Then, select the polygon feature with the **Select Feature** tool on the **Standard** toolbar.

You can clip raster datasets, mosaic datasets, or image service layers, based on the following:

- The selected polygon feature
- The selected graphic polygon that is created with the specific tool on the Draw toolbar
- The current display extent

10. First, we will clip the Pansharp_Composite_L71183029_02920070727_B10. TIF image. In the **Image Analysis** window, select the panchromatic image. Then, in the **Processing** section, click on the **Clip** button.

11. In **Table Of Contents**, turn off all layers except the newly added layer named Clip_Pansharp_Composite_L71183029_02920070727_B10.TIF.

12. Change its name to `Clip_Composite_15m`. Right-click on the image and select **Zoom To Raster Resolution**.

13. In the **Image Analysis** window, click on the **Options** button in the upper left-hand corner. In **Image Analysis Options**, make sure that the **Red Band** is set to `3` and the **Infrared Band** is set to `4`. Click on **OK**.

14. In the **Display** list, select the **Clip_Composite_15m** raster layer. Then, in the **Processing** section, click on the **NDVI** button.

15. A new raster layer named `NDVI_ Clip_Composite_15m` is added to **Table of Contents**. Change its name to `NDVI_15m`.

16. Expand and inspect its legend. When finished, collapse it. Explore the image.

17. In the **Image Analysis** window, select the **NDVI_15m** layer. Then, in the **Processing** section, click on the **Colormap to RGB** button.

18. A new raster layer named `RGB_NDVI_15m.tif` is added. In the **Image Analysis** window, select **Min-Max** for the stretch method and **Cubic Convolution** for the resampling method during the display.

    The healthy vegetation is dark green. The water ranges from yellow to orange. The bare soil, urban areas, and roads are white or range from light yellow to yellow:

19. To save the `RGB_NDVI_15m` raster layer as a permanent raster dataset in the `SatelliteImagery` folder, select it in the **Image Analysis** window and click on the **Export** button.

20. Make sure that the path fo raster **Location** is `..\LearningArcGIS\Chapter9\ SatelliteImagery`. For **Compression Type**, select **LZW**. Click on **Save**.

21. Click on **No** in the **Output Raster** message window. The NDVI raster doesn't have the `NoData` values, and the pixel depth information will increase the size of the raster dataset.

22. Click **Yes** to add the exported raster dataset to the map as a raster layer.

23. In the **Image Analysis** window, select `RGB_NDVI_15m.tiff` and set the display characteristics, as shown in the previous screenshot.

24. When finished, save your changes in a map document named `ImageProcessing.mxd` to `<drive>:\LearningArcGIS\Chapter9\ SatelliteImagery`.

25. Close ArcMap.

You can find the results at `<drive>:\LearningArcGIS\Chapter9\Results\ SatelliteImagery\ImageProcessing.mxd`.

# Summary

In this chapter, you learned how to display an aerial image on its true position on the Earth's surface in ArcMap. You explored different display techniques in ArcMap to enhance the image display for visual interpretation of the satellite imagery. You also used the **Image Analysis** toolbar to perform three basic image processing operations on the satellite imagery: building a multispectral (or composite) image, creating a high-resolution multiband image using the pan-sharpening technique, and computing and displaying NDVI for a multispectral image.

In the next chapter, you will design a poster-size map using the ArcGIS map layout.

# 10
# Designing Maps

In this chapter, we will create a presentation map by applying the basic cartographic design principles. We will add specific map elements such as graphic scale, legend, graticule, north arrow, map title, map border, and some explanatory notes. Finally, we will set up the map for printing, and will export the map in a PDF document format.

By the end of this chapter, you will have learned the following:

- Working with the ArcGIS map layout
- Organizing map elements in a layout
- Using a map template to design a presentation map

## Working with the map layout

When you create a map, you should consider the following cartographic design aspects:

- **Audience**: For example, this is a research team that performs a socio-demographic analysis
- **Purpose of the map**: For example, this is a map that shows the population density of the capital cities in Europe
- **Map scale**: For example, this is the main map body on the poster being represented at a scale of 12,000,000
- **Paper size and orientation**: For example, this is the presentation poster to be printed on a 42x59.4 cm piece of paper (A2 Portrait)
- **Map presentation**: For example, these are printed maps or posters, raster and vector files, or web maps

# Creating a map layout

The ArcMap application provides two different data views: **Data View** and **Layout View**.

The **Data View** is used when you want to add, edit, symbolize, and analyze the data. The **Layout View** allows you to design maps for print or digital display on a virtual work page named *map layout* or *page layout*. On a page layout, you can add specific map elements such as legend, graphic or numeric scale, north arrow, textual information, graph, graphic border, and the map's title. All those map elements help the audience to read, analyze, and interpret the map's message.

The map layout design is influenced by the cartographic principles. To create an effective map, you should carefully choose the datasets, the scale, the coordinate system, the symbology, and legal requirements, such as specific title, specific dates, signature lines, and more.

The main map body defines the central theme of a poster. It should be the most visually prominent map element on the virtual work page. The rest of the map elements, such as legend, graphic scale, and other data frames, should be smaller and placed in the remaining area of the map layout, as shown in the following screenshot of the `USACounties.mxd` layout:

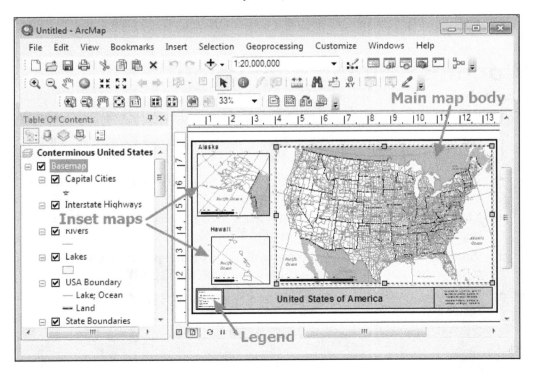

The visual hierarchy of the map elements influences the user's perception of the information transmitted by the map. The hierarchy on a map layout can be established by the position, size, color, background color, and shape of the map elements.

To preview the proposed poster-size map, please open the PDF file named `Results_CapitalCitiesByPopulation.pdf`, stored at `<drive>:\LearningArcGIS\Chapter10\Results`.

Follow these steps to start designing a map layout using the ArcMap application:

1. Start the ArcMap application, and open the existing map document `Europe.mxd` from `<drive>:\LearningArcGIS\Chapter10`.

2. Open the **Catalog** window, and connect to the folder, `Chapter10`. Click on the **Connect To Folder** button, navigate to the `<drive>:\LearningArcGIS\Chapter10` folder, and click on **OK**.

3. The map document contains a data frame named `Europe`, which has a projected coordinate reference system named `ETRS 1989 Lambert Azimuthal Equal Area`. To check the coordinate system of the data frame, right-click on the **Europe** data frame, and navigate to **Properties | Coordinate System**.

4. The `Europe` data frame displays the layers in the **Data View**. To start designing the presentation map, we will display the layers in the layout view. From the **View** menu, select **Layout View** to access the map design environment, as shown in the following screenshot:

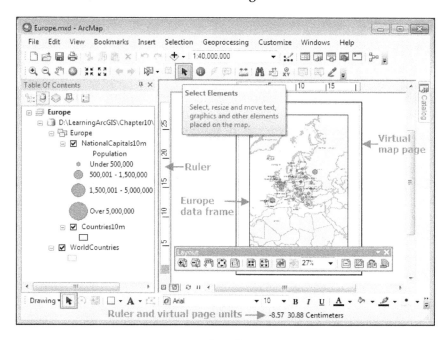

5. Notice that a new toolbar named **Layout** opens automatically.

Next, we will set the rules, rule guides, and virtual page units for the layout view:

6. From **Customize**, choose **ArcMap Options**, and select the **Layout View** tab.

7. If necessary, for **Appearance**, check the following options: **Show scroll bars**, **Show vertical guides options**, and **Show dashed line around active data frame**.

8. For the **Rules**, set **Units** to **Centimeters**. For **Smallest Division**, set 1 **cm**. For **Snap elements to**, check **Guides** and **Rules**. Make sure the **Snapping Tolerance** is set to 0.5 **cm**. Click on **OK**.

Before beginning to design a map layout, we will set the page size:

9. From the **File** menu, select **Page and Print Setup**. In the **Map Page Size** section, uncheck **Use Printer Paper Settings**. For **Standard Sizes**, select **A2**. Make sure the **Orientation** is set as **Portrait**.

 You can also get to the **Page and Print Setup** by right-clicking on the virtual work page in **Layout View**.

10. In the **Paper** section, set the **Size** to **A2**. Click on **OK**.

11. Let's save the changes in a new map document. In the **File** menu, select **Save As**. Go to the `<drive>:\LearningArcGIS\Chapter10` folder, and rename the map document as `MyEurope.mxd`.

Next, we will resize the `Europe` data frame, and change the map scale:

12. On the **Tools Toolbar**, click on the **Select Element** button, and right-click anywhere on the Europe data frame to select it; then choose **Properties**. Select the **Size and Position** tab. In the **Size** section, set **Width** to 38 **cm** and **Height** to 32 **cm**. Click on **OK** to close the **Data Frame Properties**. Notice that the values of the rules have changed.

13. On the **Standard** toolbar, type 12,000,000 in the scale box, and press the *Enter* key. In the section, *Using coordinate reference systems*, of *Chapter 2*, *Using Geographic Principles*, we explained how you can mix different units of measurement using the relative scale in ArcMap.

14. If necessary, use the **Pan** tool on the **Standard** toolbar to center the `Countries10m` layer in the data frame.

- ○ Use the **Zoom** and **Pan** tools on the **Standard** toolbar when you want to zoom or pan the data in the active data frame.

- ○ Use the **Zoom** and **Pan** tools on the **Layout** toolbar when you want to zoom or pan the virtual paper page.

To avoid changing the map scale while designing the map layout, we will lock the current scale of the Europe data frame:

15. Open **Data Frame Properties again**, and select the **Data Frame** tab. In the **Extent** drop-down list, select **Fixed Scale**. Make sure the **Scale** value is set to 12,000,000. Click on **OK**.

16. Notice that the scale box, **Zoom In**, **Zoom Out**, and **Full Extent** tools are disabled. Since the **Pan** tool doesn't change the scale value, you can still use the **Pan** tool to pan the data in the active data frame. Also, a good practice is to create one or more spatial bookmarks that capture the current extent in the map display. If you do something that accidentally changes the view, it is easy to get back to it using the bookmark.

Next, we will add two rule guides to the top ruler and five ruler guides to the side ruler for aligning the map elements on the presentation map:

17. Click on the top ruler to add the first light blue ruler guide. Then drag it to 2 centimeters. Add one more guide at 40 centimeters. On the side ruler, add guides at 56, 54, 22, 20, and 2 centimeters, as shown in the following screenshot:

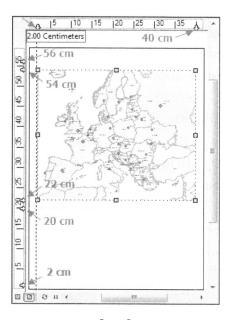

18. With the **Select Element** button active, select and drag the data frame up until it snaps to the top side ruler guide at 2 cm and to the side ruler guide at 54 cm, as shown in the preceding screenshot.

19. Save the changes to the map document.

Next, we will enhance the map by emphasizing the coastlines with multiple blue rings, as shown in the following screenshot:

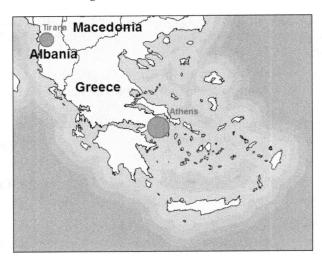

20. If necessary, open the **ArcToolbox** window. Expand **AnalysisTools | Proximity**, and double-click on the **Multiple Ring Buffer** tool. Set the following parameters:

    ○ **Input Features**: Countries10m

    ○ **Output Feature class**: ..\Chapter10\World.gdb\MultiBuffer30km

    ○ **Distances**: Type and add the values 30, 60, 90, and 120

    ○ **Buffer Unit (optional)**: Kilometers

    ○ **Field name (optional)**: distance

    ○ **Dissolve Option (optional)**: ALL

    ○ Check **Outside Polygons Only (optional)**

21. Click on **OK**. Then, right-click on **MultiBuffer30km** and navigate to **Properties | Symbology**.

22. For convenience, you will import the symbology definition from an existing layer. Under the **Show** area, navigate to **Features | Single Symbol**. Click on the **Import** button.

23. For **Layer**, click on the **Browser** button, and select `ColorRamp_MultiBuffer.lyr` from the `<drive>:\LearningArcGIS\Chapter10` folder. Click on **Add,** and then on **OK**.

24. Under **Table Of Contents**, turn off the `WorldCountries layer` (uncheck it).

Next, we will change the data frame's background color to the fourth buffer color:

25. Double-click on the **Europe** data frame to open the **Data Frame Properties** window. Select the **Frame** tab. In the **Background** section, click on the down arrow icon, and select **Blue**.

26. Click on the blue color button, and select **More Color** to change the RGB component of the blue color. For RGB values, type the following values: `122` (R), `182` (G), and `245` (B). Click on **OK**.

    Since our map will be designed only for screen display, we will use the RGB color system. If you intend to use the poster map for printing, work with the CMYK color system.

    To find out the RGB values for a certain color on the map, use the **Eye Dropper** tool. On the **Layout** toolbar, click on **Toolbar Options**, and select **Customize | Commands**. In the **Show commands containing** text box, type `eye dropper` to find the tool.

27. Inspect the results in the map layout using the **Zoom In** and **Pan** tools on the **Layout** toolbar. If necessary, use the **Refresh** button of the **Layout View** window button.

You may notice that some dynamic labels in the `NationalCapitals10m` and `Countries10m` layers overlap with the city point symbols or with other labels in the map. Next, we will adjust the position of the labels by converting them to map annotation:

28. In the **Table Of Contents section**, right-click on the **NationalCapitals10m** layer and choose **Convert Labels to Annotation**. Under **Store Annotation,** check **In the map**. Click on **Convert**.

29. Repeat the previous step for the `Countries10m` layer. If necessary, right-click on the labels in the **Overflow Annotation** window, and select **Add Annotation**. Close the empty dialog window.

> If you want to review the two annotation groups created in the last two steps and the name of the associated layers, open the **Data Frame Properties** window, and select the **Annotation Groups** tab.

30. To edit the annotations in the layout view, select the **Focus Data Frame** button in the **Layout** toolbar, as shown in the following screenshot. Notice the diagonal hatches around the "focused" data frame on the map layout.

31. With the **Select Elements** tool selected, click on each annotation text to select it. Drag it to a better position, as shown in the following screenshot:

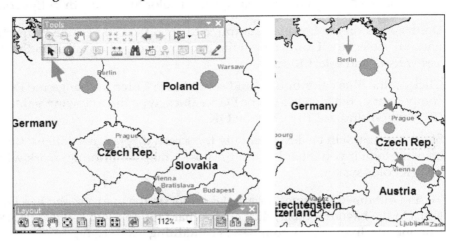

32. When finished, deselect the **Focus Data Frame** button. On the **Layout** toolbar, click on the **Zoom Whole Page** tool to return to the full extent of your map layout.

33. Save the changes in the MyEurope.mxd map document.

34. Leave the map document open to continue designing the map in the next section.

You can find the results at: `<drive>:\LearningArcGIS\Chapter10\Results\Europe_CreatingMapLayout.mxd`.

# Designing the map elements

In this section, we will create and modify a map legend, and will add and modify different map elements such as graticule, graphic scale, textual information about the map projection, and a logo image.

To organize the map elements on the virtual page, **Layout View** provides you with different alignment features such as rules, ruler guides, and grids.

All the map elements (except reference grids and graticules) are added through the **Insert** menu in the ArcMap application. To select, move, and resize a map element on a map layout, use the **Select Elements** on the **Standard** toolbar.

Follow these steps to learn how to add map elements to the map layout:

1.  Start the ArcMap application, and open your map document named MyEurope.mxd from <drive>:\LearningArcGIS\Chapter10.

First, we will add the meridians of longitude and the parallels of latitude on the map:

2.  In the map layout, right-click on the Europe data frame and select **Properties** to open its **Data Frame Properties** window. Click on the **Grid** tab. Click on the **New Grid** button.

3.  Let's explore the grid options. You can work with graticules and grids. A graticule represents the set of the meridians and parallels which cover the map area. A graticule gives you the φ, λ geographic coordinates. The option named **Graticule: divide maps by meridians and parallels** should be selected by default.

4.  A measured grid is the rectangular coordinate reference system that gives you the projected coordinates labeled X and Y or Easting (E) and Northing (N). In the **Grids and Graticules Wizard** window, check **Measured Grid: divides map into a grid of map units** to preview an example.

5.  In the **Grids and Graticules Wizard** window, check **Reference Grid: divides map into a grid for indexing** to preview an example of a map index for a city. A reference grid is a rectangular grid that divides the map area into rows and columns. The uniform blocks or grid cells can be located by the *columns*, typically labeled A, B, C, D, and so on, and *rows*, typically labeled 1, 2, 3, 4, and so on. For example, you can reference a street by the block or blocks where it lies: London Ave, A3, B3.

6.  When you've finished exploring the options, check the option named **Graticule: divide maps by meridians and parallels again**. Click on **Next**.

7.  Make sure the **Graticule and labels** option is selected. For **Intervals**, set the parallels and meridians to 10 degrees.

[  To update the graticule preview on the left, press the *Tab* key after you type 10 in the last **Deg** box next to **Place meridians every**. ]

8.  To set the color for the graticule lines and for the grid border, click on the **Style** button. Click on the **Color** button, and on the color palette, select **Gray 50%**. Click on **OK**. Go to the next panel.

9.  Under the **Labeling**, click on the **Text style** button. Change the **Size** to 16 and the Color to **Gray 50%**. Click on **OK**. Go to the next panel.

10. Under the **Graticule Border**, click on the **Line Symbol** button, and change the **Color** to Gray 50%. Click on **Finish** and **OK**.

11. Use the **Zoom** in and **Pan** tools on the **Layout** toolbar to inspect the graticule's labels. Zoom in on the upper left-hand corner of the data frame.

Next, we will simplify the labels by editing the graticule in the **Data Frame Properties** window, as shown in the following screenshot:

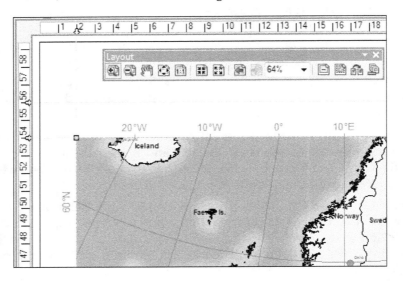

12. Reopen **Data Frame Properties | Grid**, and select **Graticule**. Click on the **Properties** button. In the **Reference System Properties** window, click on the **Labels** tab.

13. In the **Label Style** section, click on the **Additional Properties** button. In the **Minutes** and **Seconds** sections, uncheck **Show zero minutes**. Click on **OK** to close the **Grid Label Properties** window.

14. In the **Label Orientation** section, check **Left** and **Right**.

15. Next, click on the **Intervals** tab. Make sure the intervals are set to 10. In the **Origin** section, check **Use origin from the current coordinate system**. Click on **OK** to close all dialog windows.

A map showing the population density may not need a graphic scale, because its intended audience will not measure distances. However, for learning experience, we will add a scale bar for the map of Europe:

16. If necessary, zoom to the full extent of the map layout by using the **Zoom Whole Page** tool.

17. Make sure that the `Europe` data frame is selected in the map layout. From the **Insert** menu, select **Scale Bar**. In the **Scale Bar Selector** window, select **Scale Line 1 Metric**. Click on **OK**.

18. The scale bar is at the center of the data frame. A blue dotted box appears around the scale bar, which means it is already selected. Drag it to the lower left-hand corner of the **Europe** data frame. Notice that the graticule lines cross over the scale bar.

Next, we will change some of the scale bar properties, as shown in the following screenshot:

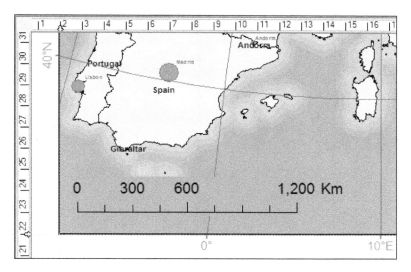

19. Right-click on the selected scale bar, and choose **Properties**. Make sure the **Scale and Units** tab is selected. In the **Scale** section, under **When resizing,** select **Adjust number of divisions**.

20. For **Division value**, type `600`. Make sure the **Division Units** is set to **Kilometers** and **Label Position** is set **after labels**. For **Label**, type `Km`.

21. We will add a color background for the scale bar to apparently cut the continuity of the graticule lines near it. It will be the color used for the Europe data frame background. Click on the **Frame** tab. In the **Background** section, click on the down arrow, and select **Blue**.

22. Click on the blue color button, and select **More Color** to change the RGB values to `122` (R), `182` (G), and `245` (B).

23. For **Gap Y**, type `6` pts. Click on **OK** to close the **Scale Line Properties** window.

Next, we will add a dynamic text which refers to the coordinate reference system of the Europe data frame:

24. On the **Layout** toolbar, click on the **Zoom Whole Page** tool.

25. Make sure the data frame is selected in the map layout. From the **Insert** menu, select **Dynamic Text | Coordinate System**. The rectangular text box is at the center of the map layout.

26. Drag it to the lower right-hand corner of the **Europe** data frame, as shown in the following screenshot:

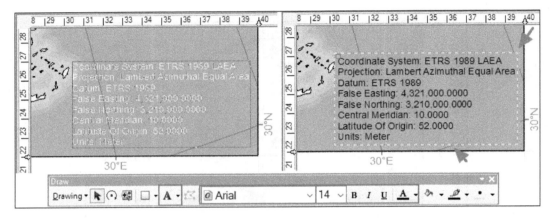

27. Right-click on the selected text box, and choose **Properties**. Click on the **Change Symbol** button. For **Size**, type 14.

You can change the text characteristics such us font, font style, size, or color, by using the specific tools on the **Draw** toolbar.

28. Click on **Edit Symbol**, and select the **Advanced Text** tab. Check **Text Background**, and click on the **Properties** button.

29. Click on the **Symbol** button, and for **Fill Color,** set the RGB values to 122 (R), 182 (G), and 245 (B). For **Outline Color**, choose **No Color**. Click on **OK** until you close all the dialog windows.

30. Since the dynamic text is linked to the data frame's properties, you cannot change the text inside the rectangular text box. If you still want to change the text, then you should transform it into a simple graphic text. Right-click on the text box, and select **Convert to Graphics**.

31. However, you can add a rectangular paragraph text using **Insert menu |
    Text**. Also, you can add a paragraph text element using **New Text** tools
    from the **Draw** toolbar, as shown in the following screenshot:

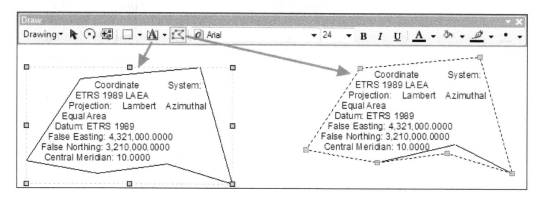

Next, we will create a legend for the Europe data frame, and will add a logo of the
www.naturalearthdata.com to emphasize the source of the data, as shown in the
following screenshot:

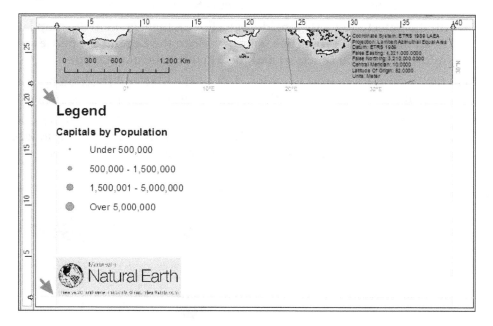

32. Select the **Europe** data frame in the map layout. From the **Insert** menu, select **Legend**. In **Legend Items**, select **Countries 10m** and **MultiBuffer30km**, and then click on the left arrow button to exclude them from the legend. A legend should only contain those layers which are specific to the purpose of the map. Layers that use common symbols do not need to be included in a legend either (Source: Tripp Corbin, 2016). This will keep the legend simple and readable.

33. Click on **Next**, and accept the defaults in the rest of the panels. Click on **Finish** to add the legend to the map layout.

34. Click on the legend, and drag it so that its upper-left corner snaps to the side ruler guide at 20 cm and to the top guide at 2 cm.

 To deselect a selected layout element, click in the empty area in the map layout.

35. Next, we will change the labels and size of the legend. In the **Table Of Contents** window, under the **NationalCapitals10m** layer name, click on the **Population heading** and change it to Capitals by Population. Notice that the change in heading is reflected in the legend.

36. With the **Select Elements** tool, double-click on the legend in the map layout to open the **Legend Properties** window. Move the **Legend Properties** window so that you can see it and the legend in the map layout.

37. In the **Legend Properties** window, click on the **Size and Position** tab. In the **Size** section, type 12 for **Width**. Click on **Apply,** and inspect the legend.

38. In the **Legend Properties** window, click on the **Items** tab. We will remove the layer name from the legend. In the **Item(s)** list, select **NationalCapitals10m**, and click on the **Style** button, as shown in the following screenshot:

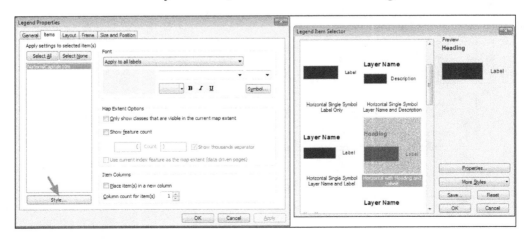

39. In the **Legend Item Selector** window, scroll down and select the **Horizontal with Heading and Labels option**. Click on **OK**.

40. In the **Legend Properties** window, click on **Apply**. Notice the changes in the the legend.

41. In the **Font** section, select **Apply to the headings** from the drop-down list. For text size, type 28. Click on **Apply,** and inspect the legend.

42. Again, in the **Font** section, select **Apply to the class labels** from the drop-down list. For text size, type 26. Click on **Apply**.

43. In the **Legend Properties** window, click on the **Layout** tab. In the **Gaps** section, you can adjust the spacing between the individual elements of the legend. Click on **OK**.

44. On the **Layout** toolbar, click on the **Zoom in** tool, and drag a box around it.

45. Since the legend size has been changed, drag the legend again so that its upper-left corner snaps to the ruler guides.

46. When finished, click on the **Zoom to 100%** tool. You can now see the legend in its print size.

If you want to make changes to the individual elements in the legend, right-click on the legend, and then select **Convert to Graphics**. The manual editing of the legend should be used in the last step of the map design, because any further changes added to the NationalCapitals10m layer will not be reflected in the legend anymore.

47. On the **Layout** toolbar, click on the **Zoom Whole Page** tool. Deselect the legend by clicking on the empty area.

48. Next, we will add a logo. From the **Insert** menu, select **Picture**. Navigate to the <drive>:\LearningArcGIS\Chapter10 folder, and select the **NaturalEarthData.com.png** file.

49. Double-click on the image to open its **Properties** window. Click on the **Size and Position** tab. In the **Size** section, type 12 for **Width**.

50. In the **Position** section, set **X** to 2 **cm** and **Y** to 2 **cm**. Make sure that the lower left-hand anchor point is selected. Click on **OK**. Inspect the results.

51. Use the **Zoom Whole Page** tool on the **Layout** toolbar to return to the full extent of your map layout. Click on the empty area to deselect the map elements on the map.

52. When finished, save your changes in the MyEurope.mxd map document.

53. Leave the map document open to continue working on the map layout in the next section.

You can find the results at `<drive>:\LearningArcGIS\Chapter10\Results\`
`Europe_AddMapElements.mxd`.

# Creating an inset map

Usually, the inset map has a smaller scale map (for example, `200,000,000`) than a main map body (for example, `10,000,000`). Also, an inset map is a smaller map body on the map layout, with a visually lower prominence. If the inset map is linked to the main map through an extent indicator, the inset map will display the extent of the main data frame on a larger area.

Follow these steps to learn how to add a new data frame (inset map):

1. If necessary, start the ArcMap application, and open your map document `MyEurope.mxd` from `<drive>:\LearningArcGIS\Chapter10`.

2. From the **Insert** menu, choose **Data Frame**. An empty data frame named **New Data Frame** is added in the map layout and also in **Table Of Contents**.

3. Next, we will change the data frame's position and size in the map layout. Right-click on the selected data frame, and choose **Properties | Size and Position**. Move the **Data Frame Properties** window so that you can see that and the new data frame in the map layout, as shown in the following screenshot:

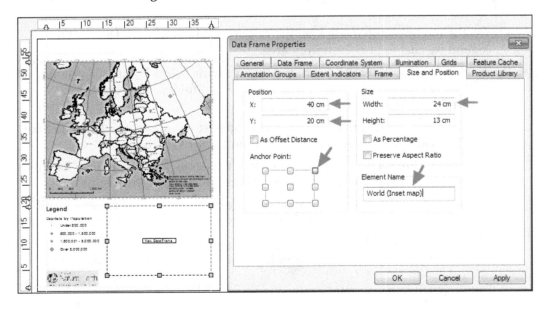

4. In the **Position** section, click on the upper right-hand anchor point to select it. Then set its position in the map layout to 40 **cm** (X) and 20 **cm** (Y).

5. In the **Size** section, set **Width** to 24 **cm** and **Height** to 13 **cm**. In the **Element Name** box, type World (Inset map). Click on **Apply** to see the changes in the map layout.

6. Click on the **Coordinate System** tab, and notice that there is no current coordinate system associated with the data frame. Click on **OK**.

Next, we will add two layers to the World data frame:

7. Under **Table Of Contents**, right-click on the WorldCountries layer in the Europe data frame, and select **Copy**. Then, right-click on the World data frame, and select **Paste Layer(s)**.

8. Turn on the **WorldCountries** layer. Open the **World** data frame **Properties** window, and notice that the geographic coordinate reference system of the WorldCountries layer named GCS_WGS_1984 is inherited automatically.

 Please notice that the name, World, of the active data frame is indicated in bold letters in **Table Of Contents**. Also, the active data frame has a gray dashed border in the map layout.

9. On the **Standard** toolbar, click on the **Add Data** button to add a base map. Navigate to the <drive>:\LearningArcGIS\Chapter10\NE1_50M_SR_W folder, and select NE1_50M_SR_W.tif. Click on **Add**. The source of the raster is NaturalEarthData.com. The raster represents a shaded relief at a scale of 1:50 million, and its symbology is based on the classes of land cover, including water.

Let's adjust the layer and the basemap transparency, as shown in the following screenshot:

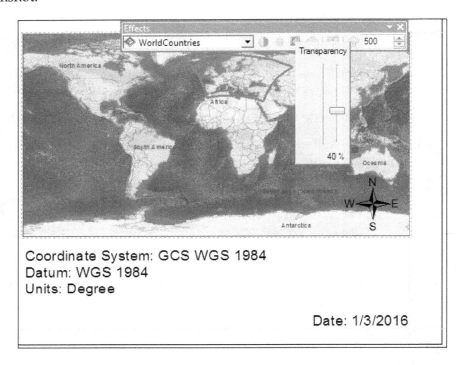

10. From the **Customize** menu, choose **Toolbars,** and check **Effects**. On the **Effects** toolbar, choose **WorldCountries** from the **Layer** dropdown list. Click on **Adjust Transparency**, and drag the slider to **40%**.

11. To have a less visually prominent background of the inset map, set the transparency of NE1_50M_SR_W.tif to **30%**.

12. On the **Tools** toolbar, click on the **Full Extent** button. Then, on the **Standard** toolbar, type 150,000,000 in the scale box and press the *Enter* key.

13. To avoid changing the map scale of the World data frame, lock the current scale as you learnt to do in the first exercise of this chapter.

Next, we will add an extent indicator on the World inset map to show the extent of the Europe data frame:

14. If necessary, select the World data frame with the **Select Elements** tool. Then right-click on the selected data frame, and choose **Properties | Extent Indicators**. Select the layer **Europe**, and click on the right arrow button to move it to the empty list on the right side.

15. In the **Options-Europe** section, click on the **Frame** button. From the **Border** drop-down list, choose **2.0 Point** border. Click on the **Color** button, and select **Mars Red**. Click on **OK** until you close all the dialog windows.

 Since the `Europe` and `World` data frames have different coordinate systems, the shape of the extent is distorted. If you change the **Coordinate System** of the `Europe` data frame to `GCS_WGS_1984`, the extent indicator gets a rectangular shape.

16. For practice, insert a north arrow and a **Coordinate System** dynamic text on the active `World` data frame. In the lower right-hand corner of the presentation map, add a **Current Date** dynamic text.

17. Try to improve the visual balance of the map by reducing the empty spaces on the virtual map page. Use the resize, move, align and group tools to refine the map layout.

Next, we will add the title of the presentation map and a border around the map contents, as shown in the following screenshot:

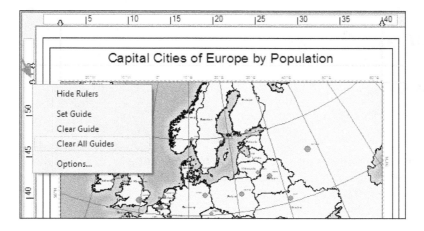

18. From the **Insert** menu, choose **Title**. In the title text box, type `Capital Cities of Europe by Population`.

19. The map's title is selected and centered in the map layout. Drag it up until it snaps to the top side ruler guide at 2 cm and to the side ruler guide at 56 cm. Then, right-click on the map's title and navigate to **Align | Align Center**.

20. From the **Insert** menu, choose **Neatline**. Check the option **Place around all elements**. From the **Border** drop-down list, choose the **Double Line** border. Make sure the **Background** is <none>. Set the **Gap** to 10 **pts**. Click on **OK**.

21. You can adjust the neatline's size and position from its **Properties** dialog window, or manually, by using the **Select Element** tool.

22. Use the **Zoom Whole Page** tool on the **Layout** toolbar to return to the full extent of your map layout. Inspect the results.

> The ruler guides are not shown on the printed map. However, when you finish designing the map, you could delete the guides by right-clicking on a ruler guide, and selecting **Clear All Guides**.

23. Save your changes in the MyEurope.mxd map document, and leave the map document open to export the map layout to a PDF file.

You can find the results at <drive>:\LearningArcGIS\Chapter10\Results\ Europe_PresentationMap.mxd.

# Exporting the Map

You can publish the maps as:

- Print maps
- File formats
- Map services

If you want to share the map with a user who doesn't have access to the ArcGIS platform, you can export the presentation map in the raster format or vector format, as shown in the following screenshot:

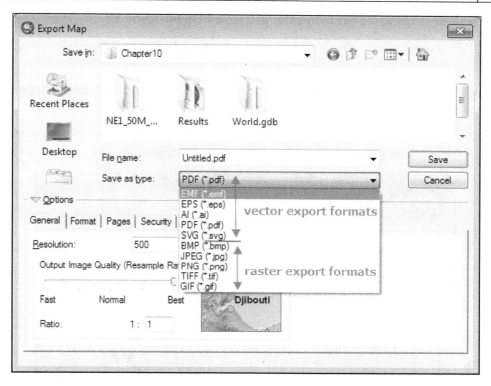

The file export format depends on the map presentation:

- **Poster printing**: for example, TIFF or AI (Adobe Illustrator allows you to choose the color system—CMYK or RGB)
- **Projected presentation**: for example, JPEG or PDF
- **Web**: for example, SVG or PNG

For more information about the map export formats, please refer to the ArcGIS Resource Center: http://resources.arcgis.com/en/help/.

Navigate to **Desktop (ArcMap): 10.4 | Map | Map export and printing**.

Follow these steps to export the poster map to a PDF file format:

1. If necessary, start the ArcMap application, and open your map document `MyEurope.mxd` from `<drive>:\LearningArcGIS\Chapter10`.

Before starting to export the presentation map, we should check whether the neatline created in the previous exercise is positioned correctly in the map layout:

2. From the **File** menu, select **Print Preview**. The neatline exceeds the printer margins, as you may see in the following screenshot:

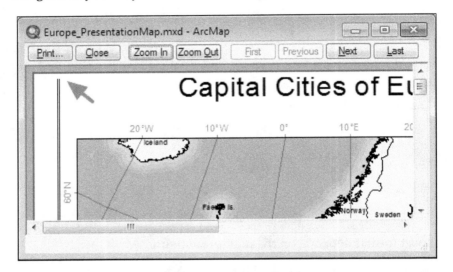

3. Click on **Close**. From the **File** menu, select **Page and Print Setup**. Check **Use Printer Paper Settings**. Make sure that the **Show Printer Margins on Layout** option is checked. Leave the **Scale Map Elements proportionally to change in Page Size** option unchecked. Click on **OK**.

Next, we will slightly adjust the neatline's size and position in the map layout, as shown in the following screenshot:

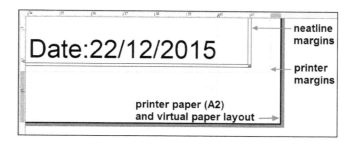

4. With the **Select Elements** tool, select the neatline. Then right-click on it, and select **Properties**.

5. Click on the **Size and Position** tab. In the **Position** section, make sure the anchor point on the lower left-hand side is selected. Then set its position in the map layout to 1 cm (X) and 1.9 cm (Y).

6. In the **Size** section, set **Width** to 40 cm and **Height** to 56 cm. Click on **OK**.

7. Inspect the results with the **Zoom in** and **Pan** tool on the **Layout** toolbar.

8. Check the neatline position again in the **Print Preview** mode. The maps should look good now.

9. Save your changes to your map document.

10. Choose **Export Map** from the **File** menu.

11. For **Save in**, navigate to `<drive>:\LearningArcGIS\Chapter10`.

12. For **Save as type**, select the **PDF (*.pdf)** format from the drop-down list.

13. In the **File Name** box, type `CapitalCitiesByPopulation`.

14. Go to **Options | General**, set the **Resolution** to 500 **dpi**. Click on the **Format** tab. For **Destination Colorspace**, check if the **RGB** is set.

15. Click on the **Save** button.

16. Explore `CapitalCitiesByPopulation.pdf` with the Adobe Reader. When done, close the ArcMap application.

You can find the results at `<drive>:\LearningArcGIS\Chapter10\Results\Results_CapitalCitiesByPopulation.pdf`.

# Working with map templates

By default, every new map document created by the ArcMap application is based on the **Blank Map** normal template that contains a single empty data frame named **Layers**, as shown in the following screenshot:

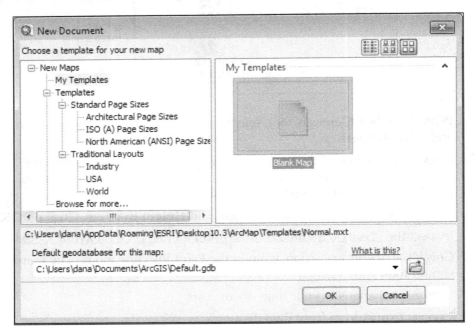

A template contains a designed map layout with one or more map bodies (data frames) and standard map elements such as legends, graphic scales, graticules, informative texts, and map title. ArcGIS for Desktop includes a lot of templates for different page sizes, and standard map layouts such as **Industry**, **World**, and **USA**, as shown in the following screenshot:

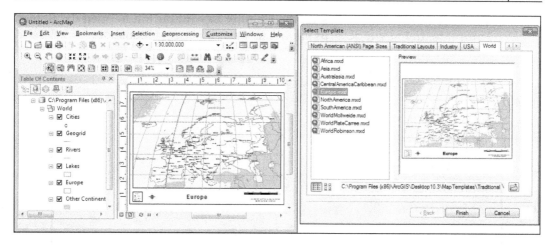

The **USA** and **World** templates include datasets. In a template, you can change the size and position of the map elements within the layout.

# Applying a map template

In this subsection, we will use the same cartographic design principles and datasets used in the section named *Working with the map layout*. We will apply a standard template named `PortraitModernInset.mxd` from the ArcMap templates collection.

To preview the final poster map, please open the PDF file named `Results_EuropeTemplate.pdf` stored in `<drive>:\LearningArcGIS\Chapter10\Results`.

Follow these steps to learn how to apply and use a map template in ArcMap:

1.  Start the ArcMap application, and open the existing map document `EuropeTemplate.mxd` from `<drive>:\LearningArcGIS\Chapter10`.

    The map document contains two data frames: `Europe` and `World`. The active data frame named `Europe` displays the layers in **Data View**.

To start designing the map, we will display the layers in the layout view:

2.  Click on the **Layout View** button on the bottom-left corner of the ArcMap window, to access the map design environment.

3.  On the **Layout** toolbar, click on the **Zoom Whole Page** button to see the full extent of the map layout.

4. On the **Layout** toolbar, click on the **Change Layout** button to open the **Select Template** dialog window, as shown in the following screenshot:

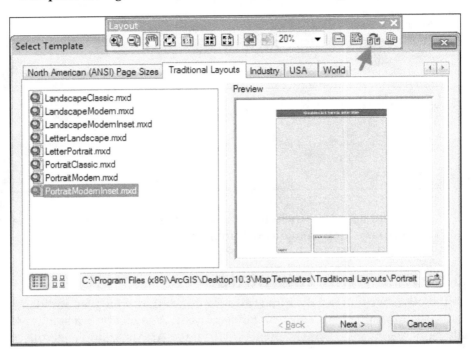

5. Click on the **Traditional Layouts** tab, and select the **PortraitModernInset.mxd** template. Click on **Next**.

6. The template includes two data frames which automatically inherit the name of the existing data frames in the EuropeTemplate map document. Make sure the main data frame (data frame 1) has the name Europe, and the inset map (data frame 2) is named World. Click on **Finish**.

7. If necessary, click on the **Zoom Whole Page** button. The active data frame named **Europe** is indicated in bold letters in **Table Of Contents**, and has a gray dashed border in the map layout.

8. Let's save the changes in a new map document. From the **File** menu, select **Save As**. Go to the <drive>:\LearningArcGIS\Chapter10 folder, and rename the map document MyEuropeTemplate.mxd.

Next, we will check the predefined page size of the map template:

9.  From the **File** menu, select **Page and Print Setup,** and move the dialog window so that you can see both that and the map layout.

10. In the **Map Page Size** section, select **Centimeters** from the **Width** and **Height** drop-down lists. The values on the layout rules have changed.

11. We will keep the ANSI C map size for our presentation map. Click on **OK**.

Next, we will set the map scale for the main data frames on the map layout:

12. In the **Table Of Contents**, right-click on the **NationalCapitals10m** layer, and select **Zoom To Layer**. Then, type 12,000,000 in the scale box, and press the *Enter* key.

13. On the **Standard** toolbar, select the **Pan** tool. Drag the map to center the **NationalCapitals10m** layer in the data frame.

14. With the **Select Elements** tool selected, right-click on the **Europe** data frame, and choose **Properties | Data Frame**. Set the **Extent to Fixed Scale**. Click on **OK**. Notice that the scale box and the **Zoom In, Zoom Out,** and **Full Extent** tools are disabled.

Next, we will resize the second data frame, and will set the map scale:

15. With the **Select Elements** tool selected, right-click on the **World** data frame, and choose **Properties | Size and Position**. Move the **Data Frame Properties** window so that you can see both that and the map layout.

16. In the **Table Of Contents section**, the World data frame is indicated in bold letters, which means it is active.

17. In the **Position** section, click on anchor point in the upper-right corner to select it. Then, in the **Size** section, set **Width** to 17 **cm**. Click on **OK,** and inspect the changes in the map layout.

18. In the **Table Of Contents**, right-click on the **WorldCountries** layer and select **Zoom To Layer**. Then, type 170,000,000 in the scale box, and press the *Enter* key. Lock the scale of the World data frame.

19. With the **Select Elements** tool selected, right-click on the blue box in the middle, and choose **Properties | Size and Position**. Set **Width** to 8 **cm** and **Height** to 11.9 **cm**. Click on **OK**. Inspect the changes in the map layout using the **Zoom In** tool on the **Layout** toolbar, as shown in the following screenshot:

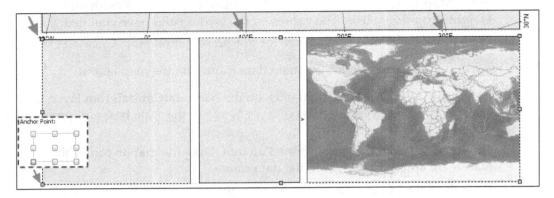

20. Save your changes to the MyEuropeTemplate.mxd map document, and leave the map document open, to continue working on the map layout in the next section.

Since you are working in the map document named MyEuropeTemplate.mxd, all previous changes in the map layout will not affect the standard template PortraitModernInset.mxd stored in <drive>:\Program Files (x86)\ArcGIS\Desktop10.4\MapTemplates\ Traditional Layouts\LandscapeModernInset.mxd.

# Modifying the map template

In this subsection, we will modify the predefined map layout by adjusting the existing map elements on it. Also, we will fill the empty sections in the map layout with additional map elements.

Follow these steps to learn how to modify the map template in the ArcMap application:

1. Start the ArcMap application, and open the existing map document MyEuropeTemplate.mxd from <drive>:\LearningArcGIS\Chapter10.

2. First, add graticule (meridians and parallels) to the Europe data frame. Use the steps learnt in the *Designing the map elements* section to edit the graticule in the **Data Frame Properties** window. Set the label size to 16 pts.

3.  The graticule's labels overlap the three blue boxes below the **Europe** main map, as shown in the following screenshot:

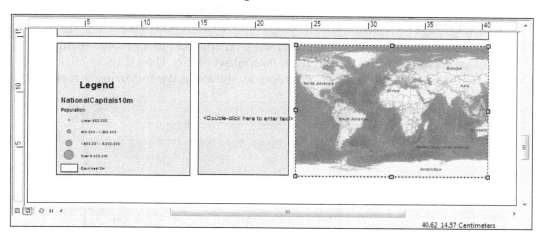

4.  A good practice is to group all three boxes to keep them aligned, and then shift them slightly on the **Y** axes. Select the three blue boxes using the **Select Elements** tool along with the *Shift* key.

5.  Right-click on the selected map elements, and choose **Group**. Then, right-click again on the group, and choose **Properties | Size and Position**. Make sure that the anchor point on the lower-left side is selected. Under **Position**, change the **Y** to 2 cm. Click on **OK**.

6.  When done, you can ungroup the three blue boxes. If you cannot see the default legend of the Europe data frame, send the selected boxes to the back using the **Order | Send to Back** tool from their context menu.

7.  With the Europe data frame selected in the map layout, add a graphic scale and the coordinate system information, as shown in the exercise from the *Designing the map elements* section.

8.  Use the **Eye Dropper** tool to find out the RGB values for the background color of the template's data frame.

9.  Next, adjust the default legend of the Europe data frame. Change the labels and size in the **Legend Properties** dialog window, as shown in the exercise from the *Designing the map elements* section.

10. Next, add an extent indicator to the World inset map to show the extent of the Europe main map body, as shown in the exercise from the *Creating an inset map* section.

11. When finished, select the **<Double-click here to enter the text>** on the middle blue box, and press the *Delete* key.

12. Optionally, reduce the middle gap in the map template to improve the visual balance of the poster. Select the **World** data frame, and insert a **Coordinate System** dynamic text. From the **Insert** menu, select **Dynamic Text | Coordinate System**.

13. To add more information, convert the dynamic text to a graphic using the **Convert to Graphics** option from its context menu. Then double-click on it, and add supplementary information, as shown in the following screenshot:

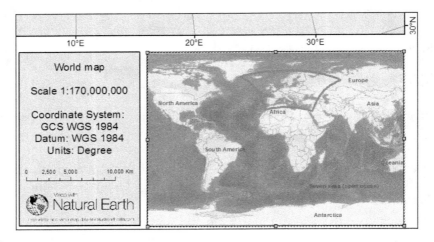

14. Insert a graphic scale and the Natural Earth logo (LearningArcGIS\ Chapter10\NaturalEarthData.com.png).

Next, change the map title at the bottom of the map template, as shown in the following screenshot:

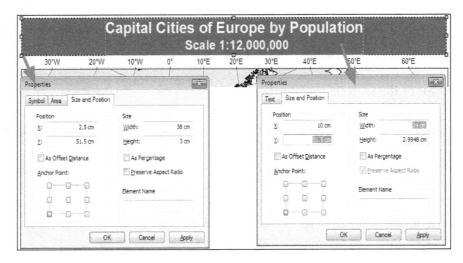

15. Select title text box **<Double-click here to enter title>** to open the **Properties** dialog window. In the **Text** box, type Capital Cities of Europe by Population.

16. Next, we will add a second text line Scale 1:12,000,000 with a small font size. To add a second text line, click on the *Enter* key.

17. We will modify the font size of the second text line by using a text formatting tag named FNT. Type <FNT size = "30"> Scale 1:12,000,000 </FNT>. If you want read more about the text formatting tags, click on **About formatting text**.

18. Click on the **Center** button to center the text. Click on **Apply**.

19. Next, select the **Size and Position** tab. In the **Position** section, set **Y** to 51.5 cm. Make sure that the anchor point on the lower-left side is selected. In the **Size** section, for **Width**, type 24 to enlarge the text box that contains the text. Click on **OK**. Inspect the results in the map layout.

20. Select the blue box behind the map title, and open the **Properties** window from its context menu. In the **Position** section, set **Y** to 51.5 **cm**. In the **Size** section, for **Width**, type 38. Click on **OK**. Inspect the results.

21. Select the blue box and the map tile again, using the *Shift* key. Right-click on the selected map elements, and navigate to **Align | Align Center**. Deselect the map elements, and zoom to the whole virtual map page.

22. When done, save the changes to the map document, and export your poster map as a PDF file.

23. Save your map document, and close the ArcMap application.

You can find the results at: <drive>:\LearningArcGIS\Chapter10\Results\EuropeTemplate.mxd.

# Summary

In this chapter, you have designed a poster-size map using the **Layout** techniques in the ArcMap application.

You have also learnt how to make a map more readable by converting labels to annotation, adjusting the transparency of the layers in a data frame, adding graticule on the map body, and adding an inset map in the map layout.

You have used and linked two data frames with different coordinate systems and different scales on a poster map.

In the second part of this chapter, you have learnt how to create a professional poster map faster and easier using a standard template from the ArcGIS collection of templates. With this, we wrap up our book!

# Index

## Symbols

**3D Analyst**
  3D features, creating from
      2D features 250-254
  TIN (Triangulated Irregular Network )
      surface, using 246-250
  using 246
**3D features**
  creating, from 2D features 250-257

## A

**additive primaries 100**
**aerial imagery**
  about 259
  orthorectified image,
      georeferencing 260-264
  using 259
**affine transformation method 151**
**ArcCatalog**
  about 11, 12
  Catalog Tree 12
  data, exploring in 13, 14
  options, for viewing selected items 12
  Preview window 12
**ArcGIS**
  attribute query 183
  location query 183
**ArcGIS 8.x 1**
**ArcGIS for Desktop**
  60-day trial, obtaining of 5-7
  about 2
  concurrent use 4
  exploring 11

hardware requisites 2, 3
  help section 23
  installing 4
  installing, on Windows 8-11
  reference link 279
  single use 4
  software requisites 2, 4
**ArcGIS Online 1**
**ArcGIS Pro**
  about 11
  URL 11
**ArcGIS Resource Center**
  reference link 267, 309
**ArcGIS Resources**
  reference link 173
**ArcMap**
  about 11-15
  Data frame 15
  Data View 16
  Layer 16
  Layout View 16
  spatial data, adding as layers to 16-21
**ArcToolbox 22, 23**
**ArcView GIS 1**
**area symbols**
  creating 110-118
**attribute data**
  creating 155
  editing 155
  feature attributes, editing 155-158
  interval 99
  nominal 99
  ordinal 99
  ratio 100

# E

# F

# G

## H

**hardware requisites, ArcGIS for Desktop**
  central processing unit (CPU) speed  2
  disk space  2
  memory/RAM  2
  Networking Hardware  3
  processor  2
  screen resolution and display properties  2
  swap space  2
  Video/Graphics Adapter  3
**horizontal datum  31, 32**
**HSV  101**
**Hue  100**

## I

**imagery  266**
**inset map**
  creating  304-308
**INSPIRE metadata standard**
  reference link  86
**intermediate data  219**
**ISO 19139 XML schema**
  reference link  94
**ISO/TC211 standards**
  reference link  86

## L

**labels**
  creating  118
  dynamic labels, working with  121-128
  graphic text, working with  118-120
  thematic map, creating  129-132
**Landsat 7 ETM+ satellite imagery  265**
**Landsat Imagery**
  reference link  265
**least-cost path analysis**
  cost of travel  240
  cost surface  240
  factors  240
  performing  240-245
  total cost surface  240
**license levels**
  advanced  231

basic  231
standards  231
**LiDAR point cloud data  234**
**line**
  creating  110-118

## M

**map**
  elements, designing  296-303
  exporting  308-311
  exporting, format  309
  exporting, reference link  309
**map document  16**
**map layout**
  creating  290-296
  inset map, creating  304-308
  map elements, designing  296-304
  working with  289
**map projections**
  azimuthal projections  54-57
  classifying  46
  comparing  47, 48
  conformal projection  46
  conic projections  51-54
  cylindrical projections  48-51
  equal Area projection  46
  equidistant projection  46
  using  46
**map template**
  applying  313-316
  modifying  316-319
  working with  312, 313
**map topology**
  creating  146-150
  editing  146-150
**metadata**
  about  13
  creating  87-91
  importing  92-96
  used, for documenting geodatabase  85-87
**ModelBuilder**
  about  219
  model, creating  220-227
  model, using  227-229
  working  219

multipatch feature  254
multipoint feature class  136
multispectral imagery
  displaying  277-281
multiuser geodatabase
  about  66
  Desktop (SQL Server Express)  66
  Enterprise  66
  Workgroup  66

# N

Natural Earth
  reference link  301
  reference links  70
NaturalEarthData.com
  data, obtaining from  13
new features
  creating  160
  digitizing  160-163
  point features, creating with
    XY data  164-166
North Pole Lambert Azimuthal Equal
  Area projection  57

# O

orthophoto
  about  259
  uses  259
orthorectified image
  georeferencing  260-264
overviews  266

# P

panchromatic band
  reference link  282
personal geodatabase  65
Plate Carrée projection  48
point features
  creating, XY data used  164-166
point symbols
  creating  101-110
polyline  135
pyramids  266

# Q

query expression  183

# R

raster data
  reference link  232
rasters  266
reference layer  260
regional datum  31, 32
relational database management
    system (RDBMS)  68
Robinson  57
root mean square error (RMS)  262

# S

satellite image interpretation
  reference link  281
satellite imagery
  about  265
  accessing  266-276
  image processing  281-287
  multispectral imagery, displaying  277-281
  using  265
snapping tolerance  136
software requisites, ArcGIS for Desktop
  admin privilege  4
  framework  4
  Internet Browser  4
  operating system  4
spatial adjustment
  using  151-155
Spatial Analyst
  least-cost path analysis,
      performing  240-245
  suitability analysis, performing  233-239
  using  232
spatial relationships, types
  adjacent to  188
  inside of  188
  intersect by  188
  within a distance of  188
sphere  26
spherical coordinate system  28, 29

spherical polar coordinate system
about 29, 30
almucantars 30
azimuth (A) coordinates 29
verticals 30
zenith distance (Z) 29
Stereographic Romania Stereo70
projection 56
sticky tolerance
reference link 143
subtractive primaries 100
suitability analysis
performing 233-239
surface area
calculating 254-257
surface feature types, Triangulated
Irregular Network (TIN) surface
breaklines 247
clip polygons 247
erase polygons 247
fill polygons 247
hard 247
replace polygons 247
soft 247
symbols
area symbols, creating 110-118
creating 99-101
line, creating 110-118
point symbols, creating 101-110

## T

thematic map
creating 129-132
three-dimensional (3D) Cartesian
coordinate system
about 30, 31
Conventional Terrestrial Pole 31
geocenter 31
Terrestrial Reference System (TRS) 31
transformation method
reference link 61
Transverse Mercator ETRS89-TM31
projection 51
Transverse Mercator ETRS89-TMzn 50

Triangulated Irregular Network (TIN)
surface
creating 246-250
reference link 246

## V

Value Attribute Table (VAT) 232
vertical datum 31, 32
volume
calculating 254-257

## W

Web Map Service (WMS) 94
weighted distance analysis. *See* least-cost
path analysis
Windows
ArcGIS for Desktop, installing on 8-11
World Geodetic System 1984 (WGS84) 31
World Mercator projection 48, 49

www.ingramcontent.com/pod-product-compliance
Lightning Source LLC
Chambersburg PA
CBHW062057050326
40690CB00016B/3125